AF691264

COMPUTER SCIENCE, TECHNOLOGY AND APPLICATIONS

FACIAL RECOGNITION TECHNOLOGY

BEST PRACTICES, FUTURE USES AND PRIVACY CONCERNS

COMPUTER SCIENCE, TECHNOLOGY AND APPLICATIONS

Additional books in this series can be found on Nova's website under the Series tab.

Additional E-books in this series can be found on Nova's website under the E-book tab.

COMPUTER SCIENCE, TECHNOLOGY AND APPLICATIONS

FACIAL RECOGNITION TECHNOLOGY

BEST PRACTICES, FUTURE USES AND PRIVACY CONCERNS

MEENA N. HARNOIS
EDITOR

LIBRARY OF CONGRESS
SURPLUS
DUPLICATE

nova publishers
New York

c.1

Copyright © 2013 by Nova Science Publishers, Inc.

All rights reserved. No part of this book may be reproduced, stored in a retrieval system or transmitted in any form or by any means: electronic, electrostatic, magnetic, tape, mechanical photocopying, recording or otherwise without the written permission of the Publisher.

For permission to use material from this book please contact us:
Telephone 631-231-7269; Fax 631-231-8175
Web Site: http://www.novapublishers.com

NOTICE TO THE READER

The Publisher has taken reasonable care in the preparation of this book, but makes no expressed or implied warranty of any kind and assumes no responsibility for any errors or omissions. No liability is assumed for incidental or consequential damages in connection with or arising out of information contained in this book. The Publisher shall not be liable for any special, consequential, or exemplary damages resulting, in whole or in part, from the readers' use of, or reliance upon, this material. Any parts of this book based on government reports are so indicated and copyright is claimed for those parts to the extent applicable to compilations of such works.

Independent verification should be sought for any data, advice or recommendations contained in this book. In addition, no responsibility is assumed by the publisher for any injury and/or damage to persons or property arising from any methods, products, instructions, ideas or otherwise contained in this publication.

This publication is designed to provide accurate and authoritative information with regard to the subject matter covered herein. It is sold with the clear understanding that the Publisher is not engaged in rendering legal or any other professional services. If legal or any other expert assistance is required, the services of a competent person should be sought. FROM A DECLARATION OF PARTICIPANTS JOINTLY ADOPTED BY A COMMITTEE OF THE AMERICAN BAR ASSOCIATION AND A COMMITTEE OF PUBLISHERS.

Additional color graphics may be available in the e-book version of this book.

Library of Congress Cataloging-in-Publication Data

ISBN: 978-1-62417-572-5

Published by Nova Science Publishers, Inc. † New York

TK
7882
.E2
F33
2013

CONTENTS

Preface		**vii**
Chapter 1	Facing Facts: Best Practices for Common Uses of Facial Recognition Technologies *Federal Trade Commission*	**1**
Chapter 2	Statement of Senator Al Franken, Chairman, Subcommittee on Privacy, Technology and the Law. Hearing on "What Facial Recognition Technology Means for Privacy and Civil Liberties"	**29**
Chapter 3	Testimony of Jennifer Lynch, Staff Attorney, Electronic Frontier Foundation. Hearing on 'What Facial Recognition Technology Means for Privacy and Civil Liberties"	**35**
Chapter 4	Statement of Nita A. Farahany, Professor of Law, Duke Law School. Hearing on "What Facial Recognition Technology Means for Privacy and Civil Liberties"	**67**
Chapter 5	Testimony of Alessandro Acquisti, Associate Professor, Carnegie Mellon University. Hearing on "What Facial Recognition Technology Means for Privacy and Civil Liberties"	**79**

Chapter 6	Statement of Brian Martin, Director of Biometric Research, MorphoTrust USA. Hearing on "What Facial Recognition Technology Means for Privacy and Civil Liberties"	**93**
Chapter 7	Testimony of Robert Sherman, Manager of Privacy and Public Policy, Facebook. Hearing on "What Facial Recognition Technology Means for Privacy and Civil Liberties"	**101**
Chapter 8	Testimony of Maneesha Mithal, Associate Director, Division of Privacy and Identity Protection, Federal Trade Commission. Hearing on "What Facial Recognition Technology Means for Privacy and Civil Liberties"	**109**
Chapter 9	Statement of Jerome M. Pender, Deputy Assistant Director, Criminal Justice Information Services Division, Federal Bureau of Investigation. Hearing on "What Facial Recognition Technology Means for Privacy and Civil Liberties"	**119**
Chapter 10	Testimony of Sheriff Larry Amerson, President, National Sheriffs' Association. Hearing on "What Facial Recognition Technology Means for Privacy and Civil Liberties"	**125**
Index		**129**

PREFACE

Having overcome the high costs and poor accuracy that once stunted its growth, one form of biometric technology - facial recognition - is quickly moving out of the realm of science fiction and into the commercial marketplace. Today, companies are deploying facial recognition technologies in a wide array of contexts, reflecting a spectrum of increasing technological sophistication. This book discusses recent and possible future advances in the use of facial recognition technologies; ways consumers can benefit from these uses; and the privacy and security concerns raised while promoting innovation.

Chapter 1 - In the 2002 film Minority Report, Steven Spielberg imagined a world in which companies use biometric technology to identify us and serve us targeted ads. Ten years later, that vision is coming closer to reality. Having overcome the high costs and poor accuracy that once stunted its growth, one form of biometric technology – facial recognition – is quickly moving out of the realm of science fiction and into the commercial marketplace. Today, companies are deploying facial recognition technologies in a wide array of contexts, reflecting a spectrum of increasing technological sophistication. At the simplest level, the technology can be used for facial detection; that is, merely to detect and locate a face in a photo. Current uses of facial detection include refining search engine results to include only those results that contain a face; locating faces in images in order to blur them; ensuring that the frame for a video chat feed actually includes a face; or developing virtual eyeglass fitting systems and virtual makeover tools that allow consumers to upload their photos online and "try on" a pair of glasses or a new hairstyle. A more refined version of facial recognition technology allows companies to assess characteristics of facial images. For instance, companies can identify moods or emotions from facial expressions to determine a player's engagement with a video game or a viewer's excitement during a movie. Companies can also place cameras into digital signs to determine the demographic characteristics of a face – such as age range and gender – and deliver targeted advertisements in real-time in retail spaces. In the most advanced application, companies can use the technology to compare individuals' facial characteristics across different images in order to identify them. In this application, an image of an individual is matched with another image of the same individual. If the face in either of the two images is identified – that is, the name of the individual is known – then, in addition to being able to demonstrate a match between two faces, the technology can be used to identify previously anonymous faces. This is the use of facial recognition that potentially raises the most serious privacy concerns because it can identify anonymous individuals in images. One prevalent current use of this application is to enable semi-automated photo tagging or photo organization on social networks and in photo management applications. On social networks this feature typically works by scanning new photos a user uploads against existing "tagged" photos. The social network then identifies the user's "friends" in the new photos so the user can tag them.

Chapter 2 – Statement of Senator Al Franken, Chairman, Subcommittee on Privacy, Technology and the Law.

Chapter 3 – Testimony of Jennifer Lynch, Staff Attorney, Electronic Frontier Foundation.

Chapter 4 – Statement of Nita A. Farahany, Professor of Law, Duke Law School.

Chapter 5 – Testimony of Alessandro Acquisti, Associate Professor, Carnegie Mellon University.

Chapter 6 – Statement of Brian Martin, Director of Biometric Research, MorphoTrust USA.

Chapter 7 – Testimony of Robert Sherman, Manager of Privacy and Public Policy, Facebook. Hearing on "What Facial Recognition Technology Means For Privacy and Civil Liberties"

Chapter 8 – Testimony of Maneesha Mithal, Associate Director, Division of Privacy and Identity Protection, Federal Trade Commission.

Chapter 9 – Statement of Jerome M. Pender, Deputy Assistant Director, Criminal Justice Information Services Division, Federal Bureau of Investigation.

Chapter 10 – Testimony of Sheriff Larry Amerson, President, National Sheriffs' Association.

In: Facial Recognition Technology ... ISBN: 978-1-62417-572-5
Editor: Meena N. Harnois © 2013 Nova Science Publishers, Inc.

Chapter 1

FACING FACTS: BEST PRACTICES FOR COMMON USES OF FACIAL RECOGNITION TECHNOLOGIES*

Federal Trade Commission

EXECUTIVE SUMMARY

"John Anderton... You could use a Guinness right about now."

– Scene from the film Minority Report depicting use of biometric technology to target individualized ads[1]

In the 2002 film Minority Report, Steven Spielberg imagined a world in which companies use biometric technology to identify us and serve us targeted ads. Ten years later, that vision is coming closer to reality. Having overcome the high costs and poor accuracy that once stunted its growth, one form of biometric technology – facial recognition – is quickly moving out of the realm of science fiction and into the commercial marketplace.

Today, companies are deploying facial recognition technologies in a wide array of contexts, reflecting a spectrum of increasing technological sophistication. At the simplest level, the technology can be used for facial detection; that is, merely to detect and locate a face in a photo. Current uses of facial detection include refining search engine results to include only those

---

* This is an edited, reformatted and augmented version of Federal Trade Commission, dated October 2012.

results that contain a face; locating faces in images in order to blur them; ensuring that the frame for a video chat feed actually includes a face; or developing virtual eyeglass fitting systems and virtual makeover tools that allow consumers to upload their photos online and "try on" a pair of glasses or a new hairstyle.

A more refined version of facial recognition technology allows companies to assess characteristics of facial images. For instance, companies can identify moods or emotions from facial expressions to determine a player's engagement with a video game or a viewer's excitement during a movie. Companies can also place cameras into digital signs to determine the demographic characteristics of a face – such as age range and gender – and deliver targeted advertisements in real-time in retail spaces.

In the most advanced application, companies can use the technology to compare individuals' facial characteristics across different images in order to identify them. In this application, an image of an individual is matched with another image of the same individual. If the face in either of the two images is identified – that is, the name of the individual is known – then, in addition to being able to demonstrate a match between two faces, the technology can be used to identify previously anonymous faces. This is the use of facial recognition that potentially raises the most serious privacy concerns because it can identify anonymous individuals in images. One prevalent current use of this application is to enable semi-automated photo tagging or photo organization on social networks and in photo management applications. On social networks this feature typically works by scanning new photos a user uploads against existing "tagged" photos. The social network then identifies the user's "friends" in the new photos so the user can tag them.

On December 8, 2011, the Federal Trade Commission ("FTC" or "Commission") hosted a workshop – "Face Facts: A Forum on Facial Recognition Technology" ("Face Facts workshop") – to explore developments in this rapidly evolving field. Panelists discussed a number of issues, including: recent advances in facial recognition technologies; current and possible future commercial uses of facial recognition technologies; ways consumers can benefit from these uses; and privacy and security concerns raised. Following the workshop, the FTC received eighty public comments discussing these issues from private citizens, industry representatives, trade groups, consumer and privacy advocates, think tanks, and members of Congress. In this report, FTC staff has synthesized those discussions and comments in order to develop recommended best practices for protecting consumer privacy in this area, while promoting innovation.

To begin, staff recommends that companies using facial recognition technologies design their services with privacy in mind, that is, by implementing "privacy by design," in a number of ways. First, companies should maintain reasonable data security protections for consumers' images and the biometric information collected from those images to enable facial recognition (for example, unique measurements such as size of features or distance between the eyes or the ears). As the increasing public availability of identified images online has been a major factor in the increasing commercial viability of facial recognition technologies, companies that store such images should consider putting protections in place that would prevent unauthorized scraping which can lead to unintended secondary uses. Second, companies should establish and maintain appropriate retention and disposal practices for the consumer images and biometric data that they collect. For example, if a consumer creates an account on a website that allows her to virtually "try on" eyeglasses, uploads photos to that website, and then later deletes her account on the website, the photos are no longer necessary and should be discarded. Third, companies should consider the sensitivity of information when developing their facial recognition products and services. For instance, companies developing digital signs equipped with cameras using facial recognition technologies should consider carefully where to place such signs and avoid placing them in sensitive areas, such as bathrooms, locker rooms, health care facilities, or places where children congregate.

Staff also recommends several ways for companies using facial recognition technologies to provide consumers with simplified choices and increase the transparency of their practices. For example, companies using digital signs capable of demographic detection – which often look no different than digital signs that do not contain cameras – should provide clear notice to consumers that the technologies are in use, before consumers come into contact with the signs. Similarly, social networks using a facial recognition feature should provide users with a clear notice – outside of a privacy policy – about how the feature works, what data it collects, and how it will use the data. Social networks should also provide consumers with (1) an easy to find, meaningful choice not to have their biometric data collected and used for facial recognition; and (2) the ability to turn off the feature at any time and delete any biometric data previously collected from their tagged photos.

Finally, there are at least two scenarios in which companies should obtain consumers' affirmative express consent before collecting or using biometric data from facial images. First, they should obtain a consumer's affirmative express consent before using a consumer's image or any biometric data derived from that image in a materially different manner than they represented when they collected the data. Second, companies should not use facial recognition to identify anonymous images of a consumer to someone who could not otherwise identify him or her, without obtaining the consumer's affirmative express consent. Consider the example of a mobile app that allows users to identify strangers in public places, such as on the street or in a bar. If such an app were to exist, a stranger could surreptitiously use the camera on his mobile phone to take a photo of an individual who is walking to work or meeting a friend for a drink and learn that individual's identity – and possibly more information, such as her address – without the individual even being aware that her photo was taken. Given the significant privacy and safety risks that such an app would raise, only consumers who have affirmatively chosen to participate in such a system should be identified.

The recommended best practices contained in this report are intended to provide guidance to commercial entities that are using or plan to use facial recognition technologies in their products and services. However, to the extent the recommended best practices go beyond existing legal requirements, they are not intended to serve as a template for law enforcement actions or regulations under laws currently enforced by the FTC. If companies consider the issues of privacy by design, meaningful choice, and transparency at this early stage, it will help ensure that this industry develops in a way that encourages companies to offer innovative new benefits to consumers and respect their privacy interests.

I. INTRODUCTION

As facial recognition technologies have become more accurate and less costly, commercial interest and investment in these technologies has grown.[2] The ability to make inferences about an individual based on his or her unique mix of facial characteristics can have countless uses, many of which are innovative and beneficial to consumers. However, the rapidly expanding commercial use of these technologies, particularly when combined with the growing availability of identified images online, can also pose complex privacy issues. Recognizing that the commercial use of these technologies will likely continue to grow, the FTC has sought to understand how they are being used, how they could be used, and the potential risks and benefits of such technologies.

The FTC's December 2011 Face Facts workshop was a first step towards exploring new developments in this field and their potential impact on consumers.[3]

The workshop featured panel discussions on the capabilities of facial recognition technologies, current and potential implementations of these technologies, and the benefits and privacy concerns these uses can generate. The workshop was followed by a one month public comment period in which Commission staff sought further input and insight on these issues.

This report builds upon the discussions at the Face Facts workshop and the written comments received thereafter to set forth a series of case studies illustrating recommended best practices for companies using or planning to use facial recognition technologies in their products or services.[4] These best practices draw upon the three core principles outlined in the FTC's March 2012 report, "Protecting Consumer Privacy in an Era of Rapid Change" ("Privacy Report").[5]

These principles are:

1. **Privacy by Design:** Companies should build in privacy at every stage of product development.
2. **Simplified Consumer Choice:** For practices that are not consistent with the context of a transaction or a consumer's relationship with a business, companies should provide consumers with choices at a relevant time and context.
3. **Transparency:** Companies should make information collection and use practices transparent.

This report begins by providing background information from the Face Facts workshop and discussing the public comments. Next, it addresses general themes that panelists and commenters raised.

Finally, it explores a series of case studies, each focused on a common commercial use of facial recognition technologies.

The recommended best practices demonstrated in the case studies are intended to provide guidance to commercial entities that are using or plan to use facial recognition technologies in their products and services. However, to the extent the recommended best practices go beyond existing legal requirements, they are not intended to serve as a template for law enforcement actions or regulations under laws currently enforced by the FTC.[6]

II. Background

A. Face Facts Workshop

Researchers, academics, industry representatives, and consumer and privacy professionals all took part in a series of wide-ranging discussions at the Face Facts workshop. The facial recognition technologies discussed included technologies that merely detect basic human facial geometry; technologies that analyze facial geometry to predict demographic characteristics, expression, or emotions; and technologies that measure unique facial biometrics.[7] Major topics included: (1) recent advances in facial recognition technologies, (2) current commercial uses of facial recognition technologies, (3) possible future uses of facial recognition technologies, and (4) privacy concerns raised by current and possible future uses of facial recognition technologies.

1. Recent Advances in Facial Recognition Technologies

Until recently, because of high costs and limited accuracy, companies have not used facial recognition technologies on a widespread basis. However, recent years have brought steady improvements in these technologies. For example, from 1993 to 2010, tests conducted by the National Institute of Standards and Technology ("NIST") showed that the false reject rate – the rate at which facial recognition systems incorrectly rejected a match between two faces that are, in fact, the same – was reduced by half every two years.[8] In 2010, in controlled tests, the error rate stood at less than one percent.[9]

Workshop panelists identified several developments that have contributed to the increased accuracy in facial recognition systems. For example, better quality digital cameras and lenses create higher quality images, from which biometric data can be more easily extracted.[10] In addition, the goal of some facial recognition technologies is to match an image of an unknown face to an identified "reference photo," where the name of the individual is known. Until recently, it was difficult to match two images if the photos were taken from different angles. With current technologies, companies can generate 3D face images to help reconcile pose variations in different images.[11]

These recent technological advances have been accompanied by rapid growth in the availability of photos online.[12] Panelists noted that ten years ago, most of the images available online were of celebrities, while today there are many sources of identified images of private citizens online.[13] One explanation for this is the rise in popularity of social networking sites. For example, in a single month in 2010, 2.5 billion photos were uploaded to Facebook.[14]

This multitude of identified images online can eliminate the need to purchase proprietary sets of identified images, thereby lowering costs and making facial recognition technologies commercially viable for a broader spectrum of commercial entities.[15]

2. Current Commercial Uses of Facial Recognition Technologies

Facial recognition technologies currently operate across a spectrum ranging from facial detection, which simply means detecting a face in an image, to individual identification, in which an image of an individual is matched with another image of the same individual. In the latter example, if the face in either of the two images is identified – *i.e.* the name of the individual pictured is known – then, in addition to demonstrating a match between two faces, the technology can be used to identify previously anonymous faces. In between these two divergent uses are a range of possibilities that include determining the demographic characteristics of a face, such as age range and gender, and recognizing emotions from facial expressions.[16]

Current uses of facial detection include, among others, refining search engine results to include only those results that contain a face, locating faces in images in order to blur or de- identify them, or ensuring that the frame for a video chat feed actually includes a face.[17] Facial detection is also used in virtual eyeglass fitting systems and virtual makeover tools that allow consumers to "try on" a pair of glasses or a new hairstyle online. In these systems, after the consumer has uploaded a photo of herself to the website, that photo is scanned, basic facial features are picked out and – using the detected facial features as reference points – the eyewear or hairstyle is superimposed on the consumer's face.[18]

More sophisticated technologies that not only distinguish a face from surrounding objects, but also assess various characteristics of that face, can be used commercially in a variety of ways. For instance, technologies that identify moods or emotions from facial expressions can be used to determine a player's engagement with a video game or a viewer's excitement during a movie.[19] Further, technologies that can determine the gender and age range of the person standing in front of a camera can be placed into digital signs or kiosks, allowing advertisers to deliver an advertisement in real-time based on the demographic of the viewer.[20] This could provide substantial benefits to advertisers by allowing them to quickly show relevant products and deals, possibly leading to more sales.[21]

One company – called SceneTap – has also leveraged the ability to capture age range and gender to determine the demographics of the clientele of bars and nightclubs.[22]

Both the operators of the venue and third parties – such as liquor distributors – can use facial data to understand the demographics of a particular venue's customers at certain times, and possibly tailor their specials or promotions accordingly.[23] SceneTap also makes the aggregate information it collects available through a mobile app that consumers can use to make decisions about which venues to patronize.[24] While these implementations do more than simply detect a face in an image, they do not derive unique biometric data for comparison purposes.

Technologies that do derive unique biometric data for comparison and identification have been implemented in a variety of manners. For example, they can be used for authentication purposes by enabling a mobile phone user to use her face, rather than a password, to unlock her phone.[25] One of the most prevalent current uses of this technology is to enable semi-automated photo tagging or photo organization on social networks and in photo management applications.[26]

As currently implemented, these features on social networks only suggest "tags" of people that the user already knows, either through a "friend" relationship or other contacts that suggest the two individuals know each other.[27]

3. Possible Future Uses of Facial Recognition Technologies

In addition to discussing current uses of facial recognition technologies, workshop panelists discussed ways in which companies could implement these technologies in the future. Most of this discussion centered around the possibility that it may become feasible to use facial recognition to identify anonymous individuals in public places, such as streets or retail stores, or in unidentified photos online.

While it does not seem that it is currently possible for commercial entities to accomplish this on a wide scale, recent studies suggest that in the near future, it may be possible. For example, in a 2011 study, Carnegie Mellon researchers were able to identify individuals in previously unidentified photos from a dating site, by using facial recognition technology to match them to their Facebook profile photos.[28]

4. Privacy Concerns Raised by Current and Possible Future Uses of Facial Recognition Technologies

As illustrated by the above examples, companies can use facial recognition technology in ways that benefit consumers by providing them innovative products and services, such as the ability to try beauty products by uploading their faces to the Web, the ability to target search results, and the ability to organize and manage photos. Companies can also use the technology to protect privacy, by, for example, detecting and blurring images in photos, or using faces instead of passwords as an authentication device to unlock mobile phones.

At the same time, the use of facial recognition technologies can raise privacy concerns. For example, panelists voiced concerns that databases of photos or biometric data may be susceptible to breaches and hacking.[29] Further, panelists discussed how some consumers may perceive digital signs equipped with cameras using facial recognition technologies as invading their privacy because they can detect consumers from a distance and process their images without their knowledge or consent.[30] Unless these signs are labeled, they often look no different to consumers than digital signs that are not equipped with cameras.

Panelists representing companies that currently use facial recognition technologies similarly acknowledged that there are privacy concerns surrounding the use of these technologies. For example, a Google representative noted the company's reluctance to implement facial recognition until it had put appropriate privacy protections in place.[31]

Perhaps of most concern, panelists surmised that advances in facial recognition technologies may end the ability of individuals to remain anonymous in public places.[32] For example, a mobile app that could, in real-time, identify anonymous individuals on the street or in a bar could cause serious privacy and physical safety concerns, although such an app might have benefits for some consumers. Further, companies could match images collected by digital signs with other information to identify customers by name and target highly-personalized ads to them based on past purchases, or other personal information available about them online.[33]

Social networks could identify non-users of the site – including children – to existing users, by comparing uploaded images against a database of identified photos. Although staff is not aware of companies currently using data in these ways, if they begin to do so, there would be significant privacy concerns.

B. Public Comments

Following the Face Facts workshop, Commission staff requested public comments regarding a number of topics and questions.[34] Among other issues, commenters were asked to provide input on: how consumers can benefit from facial recognition technologies; the privacy and security concerns surrounding the commercial use of these technologies; best practices for providing consumers with notice and choice about the use of these technologies; and best practices for deploying these technologies in a way that protects consumer privacy. The comments received reflected a wide variety of viewpoints on these issues, and are discussed below.[35]

III. General Themes

Panelists and commenters discussed several general themes that apply across the spectrum of facial recognition technologies. First, many panelists and commenters agreed that companies should implement privacy protections for all facial recognition technologies, including those that do not individually identify consumers. Commenters and panelists noted that, despite the fact that a person's face is public in the sense that numerous strangers may see it on a daily basis, it is also a persistent, unique identifier that consumers do not have the ability to replace.[36] As such, it is inextricably linked to a consumer and deserving of privacy protections.[37] FTC staff agrees with commenters who stated that companies using facial recognition technologies that operate anywhere along the spectrum – from detection to deriving unique biometric data for comparison purposes – should implement privacy protections appropriate for the context of their relationship with consumers. This conclusion is consistent with the Commission's Privacy Report, which stated that commercial entities should implement privacy protections for any data that can be reasonably linked to a specific consumer, computer, or other device.[38]

Second, a significant number of panelists and commenters agreed that it is important for companies using these technologies to increase the transparency of their data practices, given that the commercial use of the technologies is relatively new and can often be invisible to consumers.[39] To this end, panelists and commenters noted that increased consumer education about the use of facial recognition technologies is of paramount importance and that all stakeholders – including industry, trade associations, consumer and privacy groups, and government entities – should engage in consumer education efforts. FTC staff agrees that consumer education is very important. In addition to the uses and implications of facial recognition technologies, these expanded education efforts should include topics such as the potential implications of making photos publicly available.[40] This topic has particular relevance for teens, who often impulsively post photos and other content without an understanding that the photos could be used for unintended, secondary purposes. Third, several panelists and commenters discussed the role of self-regulation in this area. Participants noted that trade groups representing companies using facial recognition technologies, most notably in the digital signage industry, have proactively issued guidance and "best practices" for their members. For example, Point of Purchase Advertising International's Digital Signage Group ("POPAI") has developed a code of conduct containing recommendations for marketers to follow in order to maintain ethical data collection practices in retail settings.[41] Similarly, the Digital Signage Federation worked with the Center for Democracy and Technology to craft a voluntary set of privacy guidelines for their members, which include advertisers and digital sign operators.[42] Both of these self-regulatory codes address the use of facial recognition technologies in digital signs. FTC staff supports these efforts and encourages trade associations to explore ways to enforce compliance with their voluntary codes of conduct and privacy standards. The Commission will also continue to enforce the FTC Act against companies that engage in unfair or deceptive practices, including by failing to abide by self-regulatory programs they join. Finally, panelists and commenters discussed the need for companies implementing facial recognition to consider and implement privacy protections. In many cases, the protections suggested by panelists are covered by the principles articulated in the March 2012 Privacy Report, which are privacy by design, simplified choice, and improved transparency. To this end, the Commission staff has developed a series of case studies describing how companies can implement these principles in specific facial recognition scenarios, as described below.

IV. Case Studies on Common Commercial Uses of Facial Recognition Technologies

This report contains three case studies, each focusing on a common commercial use of facial recognition technologies. The case studies are not intended to represent an exhaustive discussion of every current use of these technologies, nor every possible use that may occur in the future. They are merely examples that demonstrate suggested best practices for common uses and represent a spectrum of increasing sophistication of the technology involved. They build on the discussions at the workshop, comments received, and principles contained in the March 2012 Privacy Report.

Case Study #1: Facial Detection

Scenario: An eyeglass company allows consumers to upload their images to the company's website and then uses facial detection to detect the face and eyes in the image and superimpose various styles of glasses on the consumer's face. The company stores the images in order to enable consumers to use this feature of the website in the future without uploading a new image.

In this scenario, the eyeglass company's use of facial detection to simply locate the consumer's face in an image the consumer has voluntarily uploaded does not, by itself, raise privacy concerns. However, there may be privacy concerns related to the company's collection and storage of the consumer's image in connection with its virtual fitting feature. The eyeglass company should take several steps to address these concerns.

First, it should design its service with privacy in mind by implementing "privacy by design." For instance, Commission staff agrees with panelists who noted that companies that hold databases of consumer images should take steps to protect those images.[43] In this example, the eyeglass company should have reasonable data security protections in place for any stored images in order to prevent unauthorized access to those images. In addition, the eyeglass company should implement a specified retention period and dispose of stored images once they are no longer necessary for the purpose for which they were collected. If a consumer deletes his or her account on the website, the stored images are no longer necessary and should be disposed of, even if the stated retention period has not yet passed.

Second, consistent with panelist and commenter recommendations on transparency and choice, the company should be clear with consumers about its data practices and provide choices appropriate for the context of the transaction. One appropriate way to achieve transparency in this scenario is to state – at the time the consumer uploads her image to try on glasses – why the company is storing, rather than immediately deleting, that image (*e.g.*, "Create an account so you can try on glasses in the future"). If the company is storing the images for a purpose that is not consistent with the context of the transaction taking place, it should provide additional information about why it is storing the images – at a "just in time" point. For example, if the company stores the images for purposes of sharing them with third parties, it should explicitly provide consumers with a choice about this practice before they upload their image – outside of a privacy policy or similar document. In all cases, the company should also inform consumers of: (1) the length of time the images are stored, (2) who will have access to the stored images, and (3) consumers' rights regarding deletion of the stored images.

Finally, consistent with the recommendations in the Commission's Privacy Report, if in the future the eyeglass company decides to use the images in a materially different manner than it represented at the time of collection, the company should obtain the affirmative express consent of the consumer prior to such use. For instance, if the eyeglass company decides to use the images in its advertising, rather than simply storing them for future use by the consumer, this would require the consumer's affirmative express consent.

Case Study #2: Detection or Recognition of Demographic Characteristics in Digital Signs

> *Scenario:* A sports drink company operates digital signs in a supermarket. These signs include cameras and have the ability to assess the age range and gender of the consumer standing in front of them. The signs display a targeted advertisement to the consumer based on those demographic characteristics.[44] The consumer's image is processed instantaneously while the consumer is standing in front of the sign and is not stored for future use.

In this scenario, the sports drink company should implement privacy by design in several ways. First, even though it is not storing images, it should still take steps to implement reasonable data security protections to protect against the possibility that a third party could hack into the sign's software and access the images in real-time. During the workshop, existing companies using facial recognition technologies for demographic detection discussed employing these types of protections. For example, SceneTap is a company that sets up cameras in bars to determine the aggregate age range and gender of a venue's patrons. It secures the feed from its cameras so that the feed cannot be accessed by anyone in the venue.[45] Staff believes that all companies operating digital signs should follow this type of approach.

Second, the sports drink company in this scenario should consider carefully where to place such signs and avoid placing them in sensitive areas. This recommendation is consistent with recommendations by both consumer groups and industry trade associations. For example, the World Privacy Forum, along with other consumer advocacy groups, drafted the Digital Signage Privacy Principles, which contain recommended consumer protection measures for digital signage using facial recognition technologies.[46] These principles state that such digital signage should not be placed in bathrooms, locker rooms, health care facilities, or areas where children congregate.[47] Similarly, POPAI's code of conduct recommends that companies not use "observed tracking data" in HIPAA-compliant areas, such as pharmacies.[48] Commission staff agrees that digital signage companies should observe these limitations, and be vigilant regarding the locations of such signs.

Third, in this scenario, the company is not storing the images. Staff considers this a best practice for companies using digital signs and other applications using similar demographic detection. As one panelist noted, because the consumer's facial characteristics are analyzed and an advertisement delivered in a matter of seconds, the underlying image quickly loses its usefulness.[49] Much of the additional data that can be derived from these digital signs, such as which demographic groups spent more time looking at particular advertisements, or how many consumers walked by a particular sign, can be stored in the form of aggregate statistics as opposed to images, thereby eliminating or at least greatly reducing the risk that the data will be identified or tied to a particular individual.[50]

With respect to choice and transparency, panelists at the Face Facts Workshop noted that consumers likely do not currently expect signs to detect their age range and gender and target an advertisement to them in real-time on the basis of these attributes.[51] Indeed, as some commenters and panelists noted, providing a clear notice is particularly important because a digital sign or kiosk that contains a camera using facial recognition technologies will often look no different to a consumer than a digital sign that does not have a camera within the display.[52] Choice is important in these situations as well. Staff agrees with commenters and panelists who noted that because the use of these technologies within digital signs is not currently consistent with reasonable consumer expectations, consumers should be offered a choice as to whether they wish to come into contact with a digital sign that markets to them based upon their perceived age range and gender.

Applying these principles to the example of the sports drink company, the company should provide clear notice that digital signs using facial recognition to detect demographic characteristics are in operation, before the consumer comes into contact with the sign. This way the consumer can choose to avoid the sign. Depending upon the size of the store, the notice may be a prominent notice at the entrance to the store itself or at the entrance to a particular section of the store – and on or near the sign itself.[53] Further, as panelists and commenters stated, a bare listing of a website or company name or logo on the notice alone would not be sufficient to inform consumers.[54] At a minimum, a notice should clearly state the purpose of the technology and indicate how consumers can find more information about the technology and the practices of the company operating the signs in that venue.[55] A consumer who does not wish to have their data used in this manner is then able to choose not to shop at this particular store or avoid the location where the sign is placed.

Panelists and commenters discussed the appropriateness of this type of "walk away" choice. Some panelists stated that it is unfair for consumers to have to avoid a retail location altogether in order to avoid these types of digital signs.[56] Other panelists suggested that the privacy risks associated with digital signs that do not store images or individually identify consumers are relatively low, and therefore this level of choice is appropriate.[57]

FTC staff supports a sliding scale approach to notice and choice. Provided that companies are sufficiently transparent, only detect age and gender, and do not store consumers' images, staff believes that a "walk away choice" may be sufficient. However, staff also encourages companies to consider other options, such as employing these technologies in kiosks that require consumer interaction prior to processing the consumer's image. For instance, Kraft Foods is reportedly considering installing face scanning kiosks in supermarkets.[58] Alternatively, companies could place digital signs only in limited sections of their stores. Staff believes that if companies begin storing consumers' images, or tracking consumers across signs, the privacy risks become much greater, and therefore the companies should provide consumers with more robust transparency and choices.

Finally, as several panelists suggested, if the sports drink company decides to begin individually identifying consumers through digital signs – such as by running those images against a database of images identified by name – the company should first obtain the consumer's affirmative express consent.[59] Obtaining the consumer's affirmative express consent is the only way to give the consumer a *meaningful* choice in this scenario, where a company is using facial recognition to identify a consumer who it could not otherwise identify. As one commenter noted, without facial recognition technology, "most individuals in public may expect that few businesses and passersby would recognize the individual's face, fewer would affix a name to the face, and fewer still would be able to associate the face with internet behavior... or other profiles."[60] The commenter further noted that, "facial recognition technology can fundamentally change that dynamic, enabling any marketer... to collect – openly or in secret – and share the identities and associated personal information of any individual whose face is captured by the camera."[61] Commission staff agrees with these observations, and it is staff's position that, before using facial recognition to identify an individual it could not otherwise identify, the company should obtain the affirmative express consent of the individual in the image.

Case Study #3: Facial Recognition in Online Social Networks

> *Scenario*: An existing social network implements a facial recognition feature. When a user uploads new photos, the social network scans those photos against existing "tagged" photos of the user's "friends."[62] The social network then identifies the user's "friends" in the new photos so the user can tag them. Users cannot utilize the feature to identify other users who are not their "friends."

The social network in this scenario should engage in privacy by design by implementing several privacy protections. First is data security, both for the images themselves as well as the biometric data derived from the images.[63] Several panelists and commenters focused on the importance of proper data security practices for companies storing biometric data.[64] One protection that has been implemented by some companies and recommended by commenters is encryption of the stored data.[65] Facebook, for example, encrypted the data that it derived from images and used to make comparisons and suggest tags.[66] Commission staff agrees that companies using facial recognition technologies should appropriately protect the images and biometric data they store.

Moreover, even if a company does not itself intend to implement facial recognition technologies, it should consider putting protections in place that would prevent unauthorized scraping of the publicly available images it stores in its online database. As discussed above, the increasing availability of identified images online has been a major factor in the increasing commercial viability of facial recognition technologies.[67] Therefore, as commenters suggested, in order to prevent an unintended secondary use of those images, companies should put in place appropriate precautions.[68]

Second, the social network described in this scenario should establish and maintain appropriate retention and disposal practices. Panelists and commenters noted that several social networking companies have implemented processes to delete biometric data when it is no longer necessary. For example, Google +'s "Find My Face" feature uses facial recognition technology to suggest tags of other users that have turned on the feature and that the user uploading the photo is connected with in some way.[69] If a user decides to turn the feature off after previously opting in, the biometric data collected about that user is deleted.[70]

In this scenario, the social network should also be transparent with consumers about its data practices regarding the facial recognition feature and provide consumers a choice about the use of facial recognition technologies.[71] Commenters and panelists noted that consumers generally provide photos and images to social networks for the purpose of organizing them and sharing them with friends.[72] While providing innovative ways to implement photo tagging may make the process of organizing and sharing photos with friends more convenient, consumers may not currently expect that their biometric data would be collected from those images and stored by the social network. Because this use is not currently within the context of consumers' relationship with the social network, when the company first rolls out this feature and begins analyzing users' photos to gather biometric data, it should provide users with a clear notice, outside of a privacy policy, about how the feature works, what data it collects, and how that data will be used. Companies should also provide consumers with an easy to find, meaningful choice not to have their biometric data collected and used for facial recognition. Further, consumers should be able to turn off the feature at any time and delete any biometric data previously collected from their tagged photos.

Finally, the social network should not collect and store biometric data of non-users of its service because there is no context in which to provide such non-users with a choice about these practices. Whereas the social network can give users of its service an opportunity to control the use of facial recognition technology, there is no practical way for the social network to offer such choice to non-users.[73] There are at least two scenarios in which social networks should obtain consumers' affirmative express consent before collecting or using biometric data from facial images.

First, as with all companies, social networks should obtain a consumer's affirmative express consent before using a consumer's image or any biometric data derived from that image in a materially different manner than it represented when it collected the data.

Second, the social network should not identify users to other users who are not their "friends" on the site without first obtaining their affirmative express consent.

Identifying an anonymous or unidentified individual in an image to someone who did not already know their identity is outside the context of the individual's relationship or transaction with the social network. In fact, at the workshop a representative of Facebook pointed out the importance of the context in which their facial recognition feature operated, stating, "we have basic principles around our use of it. Number one, it's within the social context.

So, we are not using facial recognition technology to identify people who are not known to you."[74] Offering choice on an opt-out basis would be ineffective for this use because once a person has been identified to a stranger, he or she cannot be un-identified after the fact.

A consumer's face is a persistent identifier that cannot be changed in the way that a consumer could get a new credit card number or delete a tracking cookie. Consequently, in order to offer meaningful choice to consumers about these practices, social networks should obtain consumers' affirmative express consent before identifying their anonymous images to users who could not otherwise identify them.

This last point has critical implications for the broader use of facial recognition technology. Even beyond social networks, companies should not use facial recognition to identify anonymous images of a consumer to someone who could not otherwise identify him or her, without obtaining the consumer's affirmative express consent. Consider the example of a mobile app that allows users to identify strangers in public places, such as on the street or in a bar.

If such an app were to exist, a stranger could surreptitiously use the camera on his mobile phone to take a photo of an individual who is walking to work or meeting a friend for a drink and learn that individual's identity – and possibly more information, such as her address – without the individual even being aware that her photo was taken.

Given the significant privacy and safety risks that such an app would raise, only consumers who have affirmatively chosen to participate in such a system should be identified.

CONCLUSION

The recent technological advances in the field of facial recognition undoubtedly provide a variety of benefits to consumers in the form of interesting products and services. However, as we have seen with other technologies, technological advances and the attendant business models they create often move faster than consumers' awareness or comfort. The best practices recommended in this report can guide companies as they develop new products and services and craft the processes and systems that will govern their operations. Moreover, because implementing these practices will promote consumer trust and ensure the continued growth of this industry, companies currently have incentives to engage in them. Fortunately, the commercial use of facial recognition technologies is still young. This creates a unique opportunity to ensure that as this industry grows, it does so in a way that respects the privacy interests of consumers while preserving the beneficial uses the technology has to offer. The FTC will continue to monitor this area and explore ways to work with industry, and educate and protect consumers.

DISSENTING STATEMENT OF COMMISSIONER J . THOMAS ROSCH

Facing Facts: Best Practices for Common Uses of
Facial Recognition Technologies
October 22, 2012

I respectfully dissent from the issuance of the Staff Report entitled, "Facing Facts: Best Practices for Common Uses of Facial Recognition Technologies" ("Report" or "Staff Report"). Although I appreciate Staff's efforts to examine the issues surrounding the development and use of facial recognition technology, I believe the Report goes too far, too soon. My reasoning is threefold.

First, I object to the recommendations made in the Staff Report to the extent that they are rooted in Staff's insistence that the "unfairness" prong, rather than the "deception" prong, of the consumer protection portion of Section 5 of the Federal Trade Commission Act, should govern practices relating to facial recognition technology. Section 5(n) limits our unfairness authority to an act or practice that "causes or is likely to cause substantial injury to consumers which is not reasonably avoidable by consumers themselves and not outweighed by countervailing benefits to consumers or to competition."[1] As I have pointed out before, the Commission represented in its 1980 and 1982 Statements to Congress that it will generally enforce the consumer protection "unfairness" prong of Section 5 only where there is alleged tangible injury, not simply "[e]motional impact and other more subjective types of harm."[2] The Staff Report on Facial Recognition Technology does not – at least to my satisfaction – provide a description of such "substantial injury." Although the Commission's Policy Statement on Unfairness states that "safety risks" may support a finding of unfairness,[3] there is nothing in the Staff Report that indicates that facial recognition technology is so advanced as to cause safety risks that amount to tangible injury. To the extent that Staff identifies misuses of facial recognition technology, the consumer protection "deception" prong of Section 5 – which embraces both misrepresentations and deceptive omissions – will be a more than adequate basis upon which to bring law enforcement actions.

Second, along similar lines, I disagree with the adoption of "best practices" on the ground that facial recognition *may* be misused. There is nothing to establish that this misconduct has occurred or even that it is likely to occur in the near future. It is at least premature for anyone, much less the Commission, to suggest to businesses that they should adopt as "best practices" safeguards that may be costly and inefficient against misconduct that may never occur.

Third, I disagree with the notion that companies should be required to "provide consumers with choices" whenever facial recognition is used and is "not consistent with the context of a transaction or a consumer's relationship with a business."[4] As I noted when the Commission used the same ill-defined language in its March 2012 Privacy Report, that would import an "opt-in" requirement in a broad swath of contexts.[5] In addition, as I have also pointed out before, it is difficult, if not impossible, to reliably determine "consumers' expectations" in any particular circumstance.[6]

In summary, I do not believe that such far-reaching conclusions and recommendations can be justified at this time. There is no support at all in the Staff Report for them, much less the kind of rigorous cost-benefit analysis that should be conducted before the Commission embraces such recommendations. Nor can they be justified on the ground that technological change will occur so rapidly with respect to facial recognition technology that the Commission cannot adequately keep up with it when, and if, a consumer's data security is compromised or facial recognition technology is used to build a consumer profile. On the contrary, the Commission has shown that it can and will act promptly to protect consumers when that occurs.

End Notes

[1] *Minority Report*, Dir. Steven Spielberg, DreamWorks, 20th Century Fox, 2002.

[2] Throughout this report, staff uses the term "facial recognition" to broadly refer to any technology that is used to extract data from facial images. *See* Sony, Face Recognition Technology, http://www.sony.net/SonyInfo/ technology/technology/theme/sface_01.html.

[3] FTC Workshop, *Face Facts: A Forum on Facial Recognition Technology* (Dec. 8, 2011), http://www.ftc.gov/ bcp/workshops/facefacts/. The Commission recognizes that there are many forms of biometric technology – fingerprints, retinal scans, voice-prints, etc. – that raise similar issues as facial recognition technology. However, the workshop and this report focus solely on facial recognition.

[4] This report addresses solely commercial uses and does not address the use of facial recognition technologies for security purposes or by law enforcement or government actors.

[5] FTC, Protecting Consumer Privacy in an Era of Rapid Change, Recommendations for Businesses and Policymakers, FTC Report (Mar. 2012), *available at* http://www. ftc.gov/os/2012/03/120326privacyreport.pdf. Commissioner Rosch dissented from the issuance of the Final Privacy Report. *See id.* at Appendix C.

[6] Under Section 5 of the FTC Act, the Commission is authorized to take action against unfair or deceptive acts or practices. 15 U.S.C. § 45(a). Unfair acts or practices are defined as those that cause or are likely to cause substantial injury to consumers which is not reasonably avoidable by consumers themselves and not outweighed by countervailing benefits to consumers or to competition. 15 U.S.C. § 45(n). If a company uses facial recognition technologies in a manner that is unfair under this definition, or that constitutes a deceptive act or practice, the Commission can bring an enforcement action under Section 5. In contrast, in other countries and jurisdictions, such as the European Union, in certain circumstances, consumers may need to be notified and give their consent before a company can legally use facial recognition technologies. *See Face Facts Workshop, Remarks of Simon Rice, Technology Information Commissioner's Office, United Kingdom*, at 186, 193.

[7] The biometric data derived from facial images is the unique mathematical characteristics that are extracted from the image in order to capture the individual identity. Those unique mathematical characteristics can then be compared to the characteristics extracted from other facial images to determine if there is a match. *See* Dr. Joseph. J. Atick, International Biometrics & Identification Association, *Face Recognition in the Era of Cloud and Social Media: Is it Time to Hit the Panic Button?* (Dec. 2011), at 2, *available at* http://www. ibia.org/resources.

[8] *See Face Facts Workshop, Remarks of Dr. Jonathan Phillips, NIST*, at 23-24.

[9] *See id.* These tests were done with a limited set of frontal images that were controlled for illumination; the same results could not necessarily be duplicated with snapshots taken on the street or photos posted on social networks, many of which do not contain ideal pose or lighting conditions. *See id.* at 29.

[10] *See id.* at 32.

[11] *Face Facts Workshop, Remarks of Dr. Ralph Gross, Carnegie Mellon University*, at 20.

[12] The increasing availability of identified images online is important because it allows facial recognition systems to not only match two images of the same individual, but identify that individual by name. *See also Comment of the Center for Democracy & Technology*, cmt. #87, 3.

[13] *See Face Facts Workshop, Remarks of Prof. Alessandro Acquisti, Carnegie Mellon University*, at 133, 139-140.

[14] *See id.* at 140.

[15] *See Comment of the Center for Democracy & Technology*, cmt. #87, 3; *see also Face Facts Workshop, Remarks of Dr. Ralph Gross, Carnegie Mellon University*, at 33-34 (having multiple images of a subject allows the systems to overcome difficulties such as bad lighting or a bad pose that may affect particular images).

[16] *See* Todd Bishop, *Happy or sad? You might not see that ad, if Microsoft Kinect can figure out your mood*, GeekWire, June 10, 2012, *available at* http://www.geekwire.com/2012/happy-sad-microsoft-system-target-ads- based-emotional-state; Karen Weintraub, *But How Do You Really Feel? Someday the Computer May Know*, N.Y. Times, Oct. 15, 2012, *available at* http://www.nytimes.com/2012/10/16/science/affective-programming- grows-in-effort-to-read-faces.html.

[17] *See Face Facts Workshop, Remarks of Benjamin Petrosky, Google*, at 108-110 (an image search on Google's search engine can be refined to include only image results that contain a face; its Street View service uses facial detection to blur faces that are found in its images); *Face Facts Workshop, Remarks of Gil Hirsch, face.com,* at 120-121 (face.com uses facial detection to ensure that there is actually a face in the frame for video chat feeds as a way to prevent sexually explicit video chatting). At the time of the Face Facts workshop, face.com was an independent provider of facial recognition technologies for developers. In June 2012, face.com was acquired by Facebook. Ari Levy, *Facebook Buys Face.com, Adds Facial Recognition Software*, BloomBerG, June 18, 2012, *available at* http://www. bloomberg.com/news/2012-06-18/facebook-buys-face-com-adds-facial-recognition-software.html.

[18] *See e.g.*, Ray Ban, Ray Ban Virtual Mirror, http://www.ray-ban.com/usa/science/virtual-mirror; InStyle, Hollywood Makeover, http://www.instyle.com/instyle/makeover/ 0,,,00.html.

[19] *See Face Facts Workshop, Remarks of Jai Haissman, Affective Interfaces*, at 59-60.

[20] *See Face Facts Workshop, Remarks of Brian Huseman, Intel*, at 41, 43 (Intel developed its AIM suite software that includes these technologies and has been working with large brands such as Kraft and Adidas to place it into their digital signage).

[21] A representative of Adidas noted: "If a retailer can offer the right products quickly, people are more likely to buy something." Shan Li and David Sarno, *Advertisers start using facial recognition to tailor pitches*, lA Times, Aug. 21, 2011, *available at* http:// articles.latimes.com/2011/aug/21/business/la-fi-facial-recognition-20110821.

[22] *See Face Facts Workshop, Remarks of Andrew Cummins, SceneTap*, at 66-68 (SceneTap uses Intel's AIM Suite software in its cameras to gather this demographic information).

[23] *Id.* at 67-70.

[24] *Id.*

[25] *See Face Facts Workshop, Remarks of Benjamin Petrosky, Google*, at 110-111 (Google has implemented this technology in its Face Unlock feature for Android devices).

[26] *See id.* at 111-112 (Google has enabled facial recognition technology in both its Picasa photo management software and its social network, Google+); *Face Facts Workshop, Remarks of Gil Hirsch, face.com*, at 122 (prior to its acquisition by Facebook, face.com also provided

users with the ability to tag photos on social networks); *Face Facts Workshop, Remarks of Erin Egan, Facebook,* at 222 (Facebook used facial recognition for its "Tag Suggest" feature). At the time of the Face Facts workshop, Facebook's "Tag Suggest" feature was active on the Facebook website. In 2012, Facebook suspended the feature and reportedly plans to restore it in the future. *See* Statement of Rob Sherman, Manager of Privacy and Public Policy, Facebook, *What Facial Recognition Means for Privacy and Civil Liberties: Hearing before the S. Subcomm. On Privacy, Tech. and the Law,* 112th Cong. (July 18, 2012) *available at* http://www.judiciary.senate.gov/pdf/12-7-18ShermanTestimony. pdf. As of the date of this report, the feature has not been re-activated.

[27] *See Face Facts Workshop, Remarks of Erin Egan, Facebook*, at 222; *Face Facts Workshop, Remarks of Gil Hirsch, face.com*, at 125; *Face Facts Workshop, Remarks of Benjamin Petrosky, Google*, at 153-156.

[28] This study used a limited geographic area, and therefore a limited number of photos and subjects; thus, the results cannot necessarily be duplicated on larger scale. *See Face Facts Workshop, Remarks of Prof. Alessandro Acquisti, Carnegie Mellon University*, at 130-131, 138-139. The researchers also had some success in determining the identity of students walking on a college campus by using facial recognition to match the photo taken of them walking on campus with their publicly available photos on Facebook. Further, the researchers were able to use publicly available information about the individuals they identified to indicate the likely first five digits of their social security number. These experiments used cooperative subjects willing to stop and have their picture taken; it is likely not yet possible to recreate them in order to identify anyone on the street at any time. *Id.*

[29] *See Face Facts Workshop, Remarks of Chris Conley, ACLU of Northern California*, at 166.

[30] *See Face Facts Workshop, Remarks of Fred Carter, Information and Privacy Commissioner's Office, Ontario, Canada*, at 77; *Face Facts Workshop, Remarks of Beth Givens, Privacy Rights Clearinghouse*, at 85.

[31] *See Face Facts Workshop, Remarks of Benjamin Petrosky, Google*, at 107.

[32] *See Face Facts Workshop, Remarks of Chris Conley, ACLU of Northern California*, at 149-150. Professor Acquisti also noted the possibility that this could be accomplished by incorporating facial recognition technology into mobile phone cameras, or even glasses or contact lenses. *See Face Facts Workshop, Remarks of Prof. Alessandro Acquisti, Carnegie Mellon University*, at 142-143.

[33] *See Comment of the Center for Democracy & Technology*, cmt. #87, 8 ("[I]t is likely that digital media will one day routinely identify individuals for the simple reason that it will be profitable to do so."); *see also* Stephanie Clifford, *Instant Ads Set the Pace on the Web*, N.Y. Times (March 12, 2010), *available at* http://www.nytimes. com/2010/03/12/ business/media/12adco.html (noting that the online advertising industry has been willing to pay a premium for personalized advertisements).

[34] *See* Press Release, FTC, FTC Seeks Public Comments on Facial Recognition Technology (Dec. 23, 2011), *available at* http://www.ftc.gov/opa/2011/12/facefacts.shtm.

[35] *See* FTC, FTC Seeks Public Comments on Facial Recognition Technology; Project Number P115406, http://www.ftc.gov/os/comments/facialrecognitiontechnology/index.shtm.

[36] *See Face Facts Workshop, Remarks of Prof. Alessandro Acquisti, Carnegie Mellon University*, at 135 ("Your face is a veritable conduit between your different online personas; in fact, between your offline and online persona. It's much easier to change your name and declare 'reputational bankruptcy' than to change your face. Your face creates the link between all the different personas."); *Comment of The Electronic Privacy Information Center*, cmt. #83,

12 ("unlike a credit card or social security number, it's not possible to go out and get a new faceprint if your biometric data is hacked").

[37] *See Comment of Software & Information Industry Association*, cmt. #77, 4 ("if face print information is retained, this creates a reasonable possibility of identification").

[38] There is a carve-out for entities that collect only non-sensitive data from fewer than 5,000 consumers per year and that do not share the data with third parties.

[39] *See, e.g., Face Facts Workshop, Remarks of Chris Conley, ACLU of Northern California*, at 147-149; *Face Facts Workshop, Remarks of Beth Givens, Privacy Rights Clearinghouse*, at 85; *Face Facts Workshop, Remarks of Fred Carter, Information and Privacy Commissioner's Office, Ontario, Canada*, at 77.

[40] *See Face Facts Workshop, Remarks of Erin Egan, Facebook*, at 204.

[41] *See* Press Release, POPAI, Taking Consumer Privacy Seriously, POPAI's Digital Signage Group Releases Code of Conduct (Aug. 2, 2010), *available at* http://www.popai.com/2010/02/08/taking-consumer-privacy-seriously-popais-digital-signage-group-releases-code-of-conduct/

[42] Digital Signage Federation, *Digital Signage Privacy Standards* (Feb. 2011), *available at* http://www.digitalsignagefederation.org/Resources/Documents/Articles%20and%20Whitepapers/DSF%20Digital%20Signage%20Privacy%20Standards%2002-2011%20%283%29.pdf. *Face Facts Workshop, Remarks of Harley Geiger, the Center for Democracy and Technology*, at 52.

[43] *See Face Facts Workshop, Remarks of Erin Egan, Facebook*, at 229; *see also* Dr. Joseph J. Atick, International Biometrics & Identification Association, *Face Recognition in the Era of the Cloud and Social Media: Is it Time to Hit the Panic Button* (Dec. 2011), at 5-6, *available at* http://ibia.org/resources/whitepapers/.

[44] Not all digital signs contain cameras or are equipped with facial recognition technologies. The recommendations contained within this case study are intended to apply only to digital signs that use facial recognition technologies on individuals standing in front of a camera or sensor placed in the sign.

[45] *See Face Facts Workshop, Remarks of Andrew Cummins, SceneTap*, at 71.

[46] *See Comment of World Privacy Forum*, cmt. #82, 4-5.

[47] *See id.* at 5.

[48] *See* POPAI, Digital Signage Group, *Best Practices: Recommended Code of Conduct for Consumer Tracking Research* (Feb. 2010), at 6.

[49] *See Face Facts Workshop, Remarks of Simon Rice, Technology Information Commissioner's Office, United Kingdom*, at 220-21.

[50] *See Face Facts Workshop, Remarks of Brian Huseman, Intel*, at 41, 44 (discussing the benefits aggregate statistics can provide to advertisers).

[51] *See Face Facts Workshop, Remarks of Fred Carter, Information and Privacy Commissioner's Office, Ontario, Canada*, at 77 ("you are detected whether you know it or not. And this has very important implications for applying fair information practices. You don't know it's happening. It's taken at a distance. You have no knowledge."); *Face Facts Workshop, Remarks of Harley Geiger, The Center for Democracy and Technology*, at 56 ("I have seen in press reports companies literally saying, literally declining to point out which signs actually have facial detection or facial recognition and 'say they don't want their customers to feel uncomfortable.'").

[52] *See Face Facts Workshop, Remarks of Fred Carter, Information and Privacy Commissioner's Office, Ontario, Canada*, at 77; *see also, Comment of the Center for Democracy and*

Technology, cmt. #87, 5 (noting that many digital signs using this technology are not labeled). However, at least one commenter suggested that notice may not be required when digital signs do not uniquely identify individuals. *See Comment of TechAmerica*, cmt. #79, 3. *But see Comment of the Center for Democracy and Technology*, cmt. #87, 9 (noting that transparency provides benefits to both consumers and businesses alike, because "[s]ecrecy magnifies consumers' sense that their privacy is being invaded – if companies try to hide the fact that they are using facial recognition, it will sensationalize the issue and lead consumers to more deeply distrust the technology").

[53] *See Face Facts Workshop, Remarks of Harley Geiger, Center for Democracy and Technology*, at 53-54 (suggesting layered privacy notices, beginning with a comprehensive privacy policy on the website of the owner of the device, as well as the website of the location where the camera is located; an abbreviated notice at the perimeter of the area where the technology is in use; and a short notice on the actual sign to alert consumers that the sign is using facial recognition technologies).

[54] *See Face Facts Workshop, Remarks of Pam Dixon*, at 179; *see also*, *Comment of the Center for Democracy and Technology*, cmt. #87, 16 ("While a symbol may one day alert consumers to the presence of a facial recognition or detection device or program, a symbol will only be an adequate form of notice if it is adopted on an industry- wide basis and consumers are properly educated on the meaning of the symbol").

[55] One panelist suggested that the notices in place at the venue or on the device could include a QR code that would allow consumers who wanted more in depth information to use their smart phone to obtain it. *See Face Facts Workshop, Remarks of Beth Givens, Privacy Rights Clearinghouse*, at 90-91.

[56] *See, e.g., Face Facts Workshop, Remarks of Pam Dixon, World Privacy Forum*, at 183 (stating that "we shouldn't have to live in an opt-out village"); *Face Facts Workshop, Remarks of Beth Givens, Privacy Rights Clearinghouse*, at 85-86 (stating that it is unfair for consumers to miss out on opportunities because they are uncomfortable with digital signage).

[57] *Face Facts Workshop, Remarks of Harley Geiger, Center for Democracy and Technology*, at 54 (stating that for these types of signs consumers can exercise choice by choosing not to shop in particular venues upon being notified of the practices).

[58] Shan Li and David Sarno, *Advertisers start using facial recognition to tailor pitches*, lA Times, Aug. 21, 2011, *available at* http://articles.latimes.com/2011/aug/21/business/la-fi-facial-recognition-20110821.

[59] *See, e.g., Face Facts Workshop, Remarks of Harley Geiger, Center for Democracy and Technology*, at 54-55; *Face Facts Workshop, Remarks of Dr. Joseph J. Atick, International Biometrics & Identification Association*, at 205.

[60] *Comment of the Center for Democracy and Technology*, cmt. #87, 7.

[61] *Id.*

[62] Throughout this scenario, staff uses the term "friend" to refer to an individual user that another user has a direct mutual connection with on the social network.

[63] *See United States of America (For the Federal Trade Commission) v. RockYou, Inc.*, Case No. 3:12-cv-01487- SI (Mar. 27, 2012) (consent decree), *available at* http://www.ftc.gov/os/caselist/1023120/index.shtm (the FTC alleged that RockYou failed to live up to its representation that it would take reasonable and appropriate measures to safeguard its customers personal information – including photographs – from unauthorized access).

[64] For example, one commenter raised the concern that a database containing biometric information derived from facial images could become a target for hackers and possibly lead to identity theft. *See Comment of Electronic Privacy Information Center*, cmt. #83, 12; *see*

also, e.g., Face Facts Workshop, Remarks of Chris Conley, ACLU of Northern California, at 166; *Face Facts Workshop, Remarks of Dr. Joseph J. Atick, International Biometrics & Identification Association*, at 206; *Face Facts Workshop, Remarks of Erin Egan, Facebook*, at 223.

[65] *See Comment of Electronic Privacy Information Center*, cmt. #83, 21.

[66] *See Comment of Facebook*, cmt. #81, 7.

[67] *See supra* page 4.

[68] *See* Dr. Joseph J. Atick, International Biometrics & Identification Association, *Face Detection & Face Recognition Consumer Applications: Recommendations for Responsible Use* (Dec. 2011), at 3, *available at* http://www.ibia.org/resources/ ("Protecting against building identity databases by harvesting the web requires the implementation of technical measures by those who control the repositories of social media images and search engine companies. For example, such harvesting can be prevented if the image servers block all web- crawlers that do not originate from search engine entities that have previously agreed to a declared privacy policy. Such a policy would include a commitment not to extract faceprints from these images or not to make them available for general search."); *see also, Face Facts Workshop, Remarks of Erin Egan, Facebook*, at 204 (noting that Facebook's Statement of Rights and Responsibilities prohibits unauthorized scraping).

[69] *See Face Facts Workshop, Remarks of Benjamin Petrosky, Google,* at 112-113.

[70] *See id.* at 114. Similarly, if a user opted out of Facebook's "Tag Suggest" feature, Facebook deleted any previously collected biometric data. *See Comment of Facebook*, cmt. #81, 6.

[71] *See Comment of Software & Information Industry Association*, cmt. #77, 5 (stating that the use for photo tagging on social networks should be under the control of the data subject who should be allowed to opt-in or opt-out).

[72] *See Comment of Facebook*, cmt. #81, 2; *Face Facts Workshop, Remarks of Benjamin Petrosky, Google*, at 112-114.

[73] *See Face Facts Workshop, Remarks of Gil Hirsch, face.com,* at 158-160. Further, as panelists at the workshop noted, an opt-out that requires consumers to provide photos of themselves so that a service will know not to collect their biometric data or recognize them in any images going forward would itself raise privacy concerns. *See id.*, at 158-159; *Face Facts Workshop, Remarks of John Verdi, Electronic Privacy Information Center*, at 219 (while discussing the feasibility of a "do not track" option for facial recognition, Mr. Verdi stated, "you need to be tracked in order to assert your right not to be tracked. There needs to be a generation of [biometric data] in order to compare it against the do not track database.").

[74] *See Face Facts Workshop, Remarks of Erin Egan, Facebook,* at 221.

End Notes for Dissenting Statement of Commissioner J.Thomas Rosch

[1] Federal Trade Commission Act Amendments of 1994, Pub. L. No. 103-312.

[2] *See* Letter from the FTC to Hon. Wendell Ford and Hon. John Danforth, Committee on Commerce, Science and Transportation, United States Senate, Commission Statement of Policy on the Scope of Consumer Unfairness Jurisdiction (Dec. 17, 1980), *reprinted in International Harvester Co.*, 104 F.T.C. 949, 1073 (1984) ("FTC Policy Statement on Unfairness") *available at* http://www.ftc.gov/bcp/policystmt/ad-unfair.htm; Letter from the FTC to Hon. Bob Packwood and Hon. Bob Kasten, Committee on Commerce, Science and

Transportation, United States Senate, *reprinted in* FTC Antitrust & Trade Reg. Rep. (BNA) 1055, at 568-570 ("Packwood- Kasten letter"); and 15 U.S.C. § 45(n), which codified the FTC's modern approach. Commission letter to Senators Packwood and Kastes reaffirming Statement (Mar. 5, 1982).

[3] *See supra* n.2, FTC Policy Statement on Unfairness at 3; *International Harvester Co.*, 104 F.T.C. at 1073.

[4] Report at 2.

[5] S*ee* Dissenting Statement of Commissioner J. Thomas Rosch, Issuance of Federal Trade Commission Report, *Protecting Consumer Privacy in an Era of Rapid Change: Recommendations for Businesses and Policymakers* (March 26, 2012), *available at* http://ftc.gov/speeches/rosch/120326privacyreport.pdf.

[6] *Id.*

In: Facial Recognition Technology … ISBN: 978-1-62417-572-5
Editor: Meena N. Harnois © 2013 Nova Science Publishers, Inc.

Chapter 2

STATEMENT OF SENATOR AL FRANKEN, CHAIRMAN, SUBCOMMITTEE ON PRIVACY, TECHNOLOGY AND THE LAW. HEARING ON "WHAT FACIAL RECOGNITION TECHNOLOGY MEANS FOR PRIVACY AND CIVIL LIBERTIES"*

Welcome to the fourth hearing of the Subcommittee on Privacy, Technology and the Law. Today's hearing will examine the use of facial recognition technology by the government and the private sector — and what that means for our privacy and civil liberties.

I want to be clear: there is nothing inherently right or wrong with facial recognition technology. Just like any other new and powerful technology, it is a tool that can be used for great good. But if we do not stop and carefully consider the way we use this technology, it may also be abused in ways that could threaten basic aspects of our privacy and civil liberties. I called this hearing so we can start that conversation.

I believe that we have a fundamental right to control our private information--and biometric information is already among the most sensitive of our private information, mainly because it is both unique and permanent. You can change your password. You can get a new credit card. But you can't

* This is an edited, reformatted and augmented version of testimony given on July 18, 2012 before the Senate Judiciary Committee, Subcommittee on Privacy, Technology and the Law.

change your fingerprint, and you can't change your face. Unless I guess you go to a great deal of trouble.

Indeed, the dimensions of our faces are unique to each of us—just like our fingerprints. And just like fingerprint analysis, facial recognition technology allows others to identify you with what's called a "faceprint," a unique file describing your face.

But facial recognition creates acute privacy concerns that fingerprints do not. Once someone has your fingerprint, they can dust your house or your surroundings to figure out what you've touched.

Once someone has your faceprint, they can get your name, they can find your social networking account and they can find and track you in the street, in the stores you visit, the government buildings you enter, and the photos your friends post online. Your face is a conduit to an incredible amount of information about you. And facial recognition technology can allow others to access all of that information from a distance, without your knowledge and in about as much time as it takes to snap a photo.

People think of facial recognition as something out of a science fiction movie. In reality, facial recognition technology is in broad use today. If you have a drivers' license, if you have a passport, if you are a member of a social network, chances are good that you are part of a facial recognition database.

There are countless uses of this technology, and many of them are innovative and quite useful. The State Department uses facial recognition technology to identify and stop passport fraud—preventing people from getting multiple passports under different names. Using facial recognition technology, Sheriff Larry Amerson of Alabama, who is with us here today, can make sure that a prisoner being released from the Calhoun County jail is actually the same prisoner that is supposed to be released. Similarly, some of the latest smartphones can be unlocked by the owner by just looking at the phone and blinking.

But there are uses of this technology that should give us pause.

In 2010, Facebook, the world's largest social network, began signing up all of its then-800 million users in a program called Tag Suggestions. Tag Suggestions made it easier to tag close friends in photos. That's a good thing.

But the feature did this by creating a unique faceprint for every one of those friends. And in doing so, Facebook may have created the world's largest privately-held database of faceprints-without the explicit consent of its users. To date, Tag Suggestions is an opt-out program. Unless you have taken the time to turn it off, it may have already been used to generate your faceprint.

Separately, last year, the FBI rolled out a facial recognition pilot program in Maryland, Michigan and Hawaii that will soon expand to three more states. This pilot lets officers in the field take a photo of someone and compare it to a federal database of criminal mugshots. The pilot can also help ID a suspect in a photo from an actual crime. Already, several other states are setting up their own facial recognition systems independently of the FBI. These efforts will catch criminals: they already have.

Now many of you may be thinking that that's an excellent thing. I agree. But unless law enforcement facial recognition programs are deployed in a very careful manner, I fear that these gains could eventually come at a high cost to our civil liberties.

I fear that the FBI pilot could be abused to not only identify protesters at political events and rallies, but to target them for selective jailing and prosecution, stifling their First Amendment rights. Curiously enough, a lot of the presentations on this technology by the Department of Justice show it being used on people attending political events or other public gatherings.

I also fear that without further protections, facial recognition technology could be used on unsuspecting civilians innocent of any crime — invading their privacy and exposing them to potential false identifications.

Since 2010, the National Institute of Justice, which is a part of DOJ, has spent $1.4 million to develop facial recognition-enhanced binoculars that can be used to identify people at a distance and in crowds. It seems easy to envision facial recognition technology being used on innocent civilians when all an officer has to do is look at them through his binoculars.

But facial recognition technology has reached a point where it is not limited to law enforcement and multi-billion dollar companies: it can also be used by private citizens. Last year, Professor Alessandro Acquisti of Carnegie Mellon University, who is testifying today, used a consumer-grade digital camera and off-the-shelf facial recognition software to identify one out of three students walking across a campus.

I called this hearing to raise awareness about the fact that facial recognition already exists right here, today, and we need to think about what that means for our society. I also called this hearing to call attention to the fact that our federal privacy laws are almost totally unprepared to deal with this technology.

Unlike what we have in place for wiretaps and other surveillance devices, there is no law regulating law enforcement use of facial recognition technology. And current Fourth Amendment case law generally says that we have no reasonable expectation of privacy in what we voluntarily expose to the public — yet we can hardly leave our houses in the morning without exposing our faces to the public. So law enforcement doesn't need a warrant to use this technology on someone. It might not even need to have a reasonable suspicion that the subject has been involved in a crime.

The situation for the private sector is similar. Federal law provides some protection against true bad actors that promise one thing yet do another. But that's pretty much as far as the law goes. If a store wants to take a photo of your face when you walk in and generate a faceprint — without your permission — they can do that. They might even be able to sell it to third parties.

Thankfully, we have a little time to do better. While this technology will in a matter of time be at a place where it can be used quickly and reliably to identify a stranger, it isn't there just yet. And so I have called the FBI and Facebook here today to challenge them to use their position as leaders in their fields to set an example for others—before this technology is used pervasively.

The FBI already has some privacy safeguards in place. But I still think that they could do more to prevent this technology from being used to identify and target people engaging in political protests or other free speech. I think the FBI could do more to make sure that officers use this technology only when they have good reason to think that someone is involved in a crime. I also think that if the FBI did these things, law enforcement agencies around the country would follow.

For their part, Facebook allows people to use Tag Suggestions only on their close friends. But I think Facebook could still do more to explain to its users how it uses facial recognition — and to give them better choices about whether or not to participate in Tag Suggestions. I think that Facebook could make clear to its users just how much data it has—and how it will and will not use its large and growing database of faceprints. And I think that if Facebook did these things, they would establish a best practice against which other social networks would be measured.

My understanding is that for the past few months, Facebook Tag Suggestions has been temporarily disabled to allow for some technical maintenance. It seems to me that Facebook has the perfect opportunity to make changes to its facial recognition program when it brings Tag Suggestions back online.

I'm also calling the Federal Trade Commission to testify because they are in the process of actually writing best practices for the use of this technology in industry. I urge the Commission to use this as an opportunity to guarantee consumers the information and choices they need to make informed decisions about their privacy.

In the end, though, I also think that Congress may need to act — and it wouldn't be the first time it did. In the era of J. Edgar Hoover, wiretaps were used freely with little regard to privacy. Under some Supreme Court precedents of that era, as long as the wiretapping device did not actually penetrate the person's home or property, it was deemed constitutionally sound — even without a warrant. And so in 1968, Congress passed the Wiretap Act. Thanks to that law, wiretaps are still used to stop violent and serious crimes. But police need a warrant before they get a wiretap. And you can't wiretap someone just because they're a few days late on their taxes — wiretaps can be used only for certain categories of serious crimes.

I think that we need to ask ourselves whether Congress is in a similar position today as it was 50 or 60 years ago—before passage of the Wiretap Act. I hope the witnesses today will help us consider this and all of the different questions raised by this technology.

In: Facial Recognition Technology … ISBN: 978-1-62417-572-5
Editor: Meena N. Harnois © 2013 Nova Science Publishers, Inc.

Chapter 3

TESTIMONY OF JENNIFER LYNCH, STAFF ATTORNEY, ELECTRONIC FRONTIER FOUNDATION. HEARING ON "WHAT FACIAL RECOGNITION TECHNOLOGY MEANS FOR PRIVACY AND CIVIL LIBERTIES"*

Mr. Chairman and Members of the Subcommittee:

Thank you very much for the opportunity to testify today on facial recognition. My name is Jennifer Lynch, and I am an attorney with the Electronic Frontier Foundation (EFF) in San Francisco. For the last few years, first at the Samuelson Law, Technology & Public Policy Clinic at Berkeley Law School and then at EFF, I have studied the privacy implications of new technologies, including facial recognition. I have written and presented on federal, state and local law enforcement efforts to expand biometrics databases by adding facial recognition capabilities and on the impact that would have on all Americans and especially on immigrant communities. At EFF I file and litigate Freedom of Information Act requests, including several related to biometrics and facial recognition, and analyze and report on the records I receive. I have been interviewed for and quoted on biometrics and other privacy-threatening technologies in mainstream and technical press including

* This is an edited, reformatted and augmented version of testimony given on July 18, 2012 before the Senate Judiciary Committee, Subcommittee on Privacy, Technology and the Law.

the *New York Times*, *The Economist, Los Angeles Times*, *Wall Street Journal, NPR, Wired, Huffington Post*, *CNet*, *Forbes*, and elsewhere.

Although the collection of biometrics—including face recognition-ready photographs— seems like science fiction; it is already a well-established part of our lives in the United States. The Federal Bureau of Investigation (FBI) and Department of Homeland Security (DHS) each have the largest biometrics databases in the world,[1] and both agencies are working to add extensive facial recognition capabilities. The FBI has partnered with several states to collect face recognition-ready photographs of all suspects arrested and booked,[2] and, in December 2011, DHS signed a $71 million dollar contract with Accenture to incorporate facial recognition and allow real-time biometrics sharing with the Department of Justice (DOJ) and Department of Defense (DOD). State and local law- enforcement agencies are also adopting and expanding their own biometrics databases to incorporate face recognition, and are using handheld mobile devices to allow biometrics collection in "the field."[3]

The scope of government-driven biometrics data collection is well-matched by private- sector collection. Facebook, which uses face recognition by default to scan all photos uploaded to its site, states that its users uploaded more than 300 million photos *every day* in the three months ending on March 31, 2012.[4] And Face.com, which developed Facebook's face recognition tools and was recently acquired by the company, stated in March that it had indexed 31 billion face images.[5] Other companies, from large technology companies like Google and Apple to small smartphone app providers, also provide face recognition products to their customers, and private companies are using biometric identification for everything from preventing unauthorized access to computers and corporate facilities to preventing unauthorized access to the gym.[6]

Face recognition is here to stay, and, though many Americans may not realize it, they are already in a face recognition database. Facebook refuses to say how many face prints it has in its database and whether it creates a face print for photos of non-Facebook users.[7]

However, given that Facebook has approximately 170 million active monthly users in the United States alone, at least 54% of the United States population already has a face print.[8]

Face recognition technology, like other biometrics programs that collect, store, share and combine sensitive and unique data poses critical threats to privacy and civil liberties. Biometrics in general are immutable, readily accessible, individuating and can be highly prejudicial. Face recognition, though, takes the risks inherent in other biometrics to a new level because Americans cannot take precautions to prevent the collection of their image.

Face recognition allows for covert, remote and mass capture and identification of images—and the photos that may end up in a database include not just a person's face but also how she is dressed and possibly whom she is with. This creates threats to free association and free expression not evident in other biometrics.

Americans cannot participate in society without exposing their faces to public view. Similarly, connecting with friends, family and the broader world through social media has quickly become a daily (and some would say necessary) experience for Americans of all ages. Though face recognition implicates important First and Fourth Amendment values, it is unclear whether the Constitution would protect against over-collection. Without legal protections in place, it could be relatively easy for the government or private companies to amass a database of images on all Americans.

This presents opportunities for Congress to develop legislation that would protect Americans from inappropriate and excessive biometrics collection. The Constitution creates a baseline, but Congress can legislate significant additional privacy protections. As I discuss further below, Congress could use statutes like the Wiretap Act[9] and the Video Privacy Protection Act[10] as models for this legislation. Both were passed in direct response to privacy threats posed by new technologies and each includes meaningful limits and protections to guard against over-collection, retention and misuse of data.

My testimony will discuss some of the larger current and proposed facial recognition collection programs and their implications for privacy and civil liberties in a democratic society. It will also review some of the laws that may govern biometrics collection and will outline best practices for developing effective and responsible biometrics programs—and legislation to regulate those programs—in the future.

GOVERNMENT USE OF FACIAL RECOGNITION TECHNOLOGIES

Law Enforcement and government at all levels in the United States regularly collect biometrics; combine them with biographic data such as name, address, immigration status, criminal record, gender and race; store them in databases accessible to many different entities; and share them with other agencies and governments. These collection programs have, in the past, typically included only one biometric identifier (generally a fingerprint or DNA). However, many are rapidly expanding to include facial recognition-ready photographs.

Federal and State Biometrics Databases

The two largest biometrics databases in the world are the FBI's Integrated Automated Fingerprint System (IAFIS) and DHS's Automated Biometric Identification System (IDENT), a part of its U.S. Visitor and Immigration Status Indicator Technology (US- VISIT) program.[11] Each database holds more than 100 million records—more than one third the population of the United States. Although each of these databases currently relies on fingerprints, both are in the process of incorporating facial recognition.

IAFIS's criminal file includes records on people arrested at the local, state, and federal level and latent prints taken from crime scenes. IAFIS's civil file stores biometric and biographic data collected from members of the military, federal employees and as part of a background check for many types of jobs, such as childcare workers, law-enforcement officers, and lawyers.[12] IAFIS includes over 71 million subjects in the criminal master file and more than 33 million civil fingerprints,[13] and supports over 18,000 law- enforcement agencies at the state, local, tribal, federal, and international level.

IDENT stores biometric and biographical data for individuals who interact with the various agencies under the DHS umbrella, including Immigration and Customs Enforcement (ICE), U.S. Citizenship and Immigration Services (USCIS), Customs and Border Protection (CBP), the Transportation Security Administration (TSA), the U.S. Coast Guard, and others.[14] Through US-VISIT, DHS collects fingerprints from all international travelers to the United States who do not hold U.S. passports.[15]

USCIS also collects fingerprints from citizenship applicants and all individuals seeking to study, live, or work in the United States.[16] And the State Department transmits fingerprints to IDENT from all visa applicants.[17] IDENT processes more than 300,000 "encounters" every day and has 130 million records on file.[18]

In addition to the federal databases, each of the states has its own biometrics databases, and some larger metropolitan areas like Los Angeles also have regional databases. The prints entered into these databases are shared with the FBI, and under the Secure Communities program, with DHS.

Incorporating Face Recognition Capabilities into Existing Government Databases

In the last few years, federal, state and local governments have been pushing to develop "multimodal" biometric systems that collect and combine two or more biometrics (for example, photographs and fingerprints[19]), arguing that collecting multiple biometrics from each subject will make identification systems more accurate.[20] The FBI's Next Generation Identification (NGI) database represents the most robust effort to introduce and streamline multimodal biometrics collection. FBI has stated it needs "to collect as much biometric data as possible . . . and to make this information accessible to all levels of law enforcement, including International agencies."[21] Accordingly, it has been working "aggressively to build biometric databases that are comprehensive and international in scope."[22]

The biggest and perhaps most controversial change brought about by NGI will be the addition of face-recognition ready photographs.[23] The FBI has already started collecting such photographs through a pilot program with a handful of states.[24] Unlike traditional mug shots, the new NGI photos may be taken from any angle and may include close-ups of scars, marks and tattoos.[25] They may come from public and private sources, including from private security cameras, and may or may not be linked to a specific person's record (for example, NGI may include crowd photos in which many subjects may not be identified). NGI will allow law enforcement, correctional facilities, and criminal justice agencies at the local, state, federal, and international level to submit and access photos, and will allow them to submit photos in bulk.

The FBI has stated that a future goal of NGI is to allow law-enforcement agencies to identify subjects in "public datasets," which could include publicly available photographs, such as those posted on Facebook or elsewhere on the Internet.[26] Although a 2008 FBI Privacy Impact Assessment (PIA) stated that the NGI/IAFIS photo database does not collect information from "commercial data aggregators," the PIA acknowledged this information could be collected and added to the database by other NGI users such as state and local law-enforcement agencies.[27] The FBI has also stated that it hopes to be able to use NGI to track people as they move from one location to another.[28]

Another big change in NGI will be the addition of non-criminal photos. If someone applies for any type of job that requires fingerprinting or a background check, his potential employer could require him to submit a photo to the FBI. And, as the 2008 FBI PIA notes, "expanding the photo capability within the NGI [Interstate Photo System] will also expand the searchable photos that are currently maintained in the repository." Although noncriminal information has always been kept separate from criminal, the FBI is currently developing a "master name" system that will link criminal and civil data and will allow a single search query to access all data. The Bureau has stated that it believes that electronic bulk searching of civil records would be "desirable."[29]

DHS is poised to expand IDENT to include face recognition, which would further increase data sharing between DHS and DOJ through Secure Communities and between both agencies and DOD through other programs.[30] DHS has not yet released a Privacy Impact Assessment discussing this change.

Technological Advancements Will Make Face Recognition More Prevalent

Recent advancements in camera and surveillance technology over the last few years support law enforcement goals to use face recognition to track Americans. For example, the National Institute of Justice has developed a 3D binocular and camera that allows realtime facial acquisition and recognition at 1000 meters.[31] The tool wirelessly transmits images to a server, which searches them against a photo database and identifies the photo's subject. As of 2010, these binoculars were already in field-testing with the Los Angeles Sheriff's Department. Presumably, the back-end technology for these binoculars could be incorporated into other tools like body-mounted video cameras or the MORIS (Mobile Offender Recognition and Information System) iPhone add-on that some police officers are already using.[32]

Private security cameras and the cameras already in use by police departments have also advanced. They are more capable of capturing the details and facial features necessary to support facial recognition-based searches, and the software supporting them allows photo manipulation that can improve the chances of matching a photo to a person already in the database. For example, Gigapixel technology, which creates a panorama photo of many megapixel images stitched together (like those taken by security cameras), allows anyone viewing the photo to drill down to see and tag faces from even the largest crowd photos.[33]

It also shows not just a face but also what that person is wearing; what books and political or religious materials he is carrying; and whom he is with. And image enhancement software, already in use by some local law enforcement, can adjust photos "taken in the wild"[34] so they work better with facial recognition searches.

Cameras are also being incorporated into more and more devices that are capable of tracking Americans and that can provide that data to law enforcement.

For example, one of the largest manufacturers of highway toll collection systems filed a patent application in 2011 to incorporate cameras into the transponder that sits on the dashboard in your car.[35] This manufacturer's transponders are already in 22 million cars, and law enforcement already uses this data to track subjects. While a patent application does not mean the company is currently manufacturing or trying to sell the devices, it certainly shows it's interested.

Interoperability and Data Sharing

Before September 11, 2001, the federal government had many policies and practices in place to silo data and information within each agency. Since that time the government has enacted several measures that allow—and in many cases require—information sharing within and among federal intelligence and federal, state, and local law-enforcement agencies.[36] For example, currently the FBI, DHS, and Department of Defense's biometrics databases are interoperable, which means the systems can easily share and exchange data.[37] This has allowed information sharing between FBI and DHS under ICE's Secure Communities program.[38]

And states are collecting and sharing biometric data with the federal government as well. At least 31 states have already started using some form of facial recognition with their DMV photos,[39] generally to stop fraud and identity theft, and the FBI has already worked with North Carolina, one of a handful of states reported to be in the NGI pilot program, to track criminals using the state's DMV records.[40]

States also share fingerprints (and face prints soon) indirectly with DHS through Secure Communities. According to the FBI, NGI will allow all states to share and access face prints as easily as they now share and access fingerprints by 2014.[41]

The federal government also exchanges biometric data with foreign governments through direct and ad-hoc access to criminal and terrorist databases.[42] And ICE and the FBI share biometric data on deportees with the countries to which they are deported.[43]

PRIVATE SECTOR USE OF FACIAL RECOGNITION TECHNOLOGIES[44]

Private sector use of facial recognition has expanded exponentially in the last few years as well. Facebook uses face recognition for each of its 900 million users.[45] Google offers face recognition to its 170 million Google+ users,[46] and Google and Apple both provide face recognition capabilities in their photo editing systems.[47] App developers offer face recognition to unlock a phone[48] or make tagging easier,[49] and software and hardware developers and manufacturers offer face recognition systems to identify users, and prevent unauthorized access to documents, computers and facilities.

Due to the large number of Facebook users and the fact that these users actively tag each other and themselves in photos, Facebook's face recognition system is the most robust and well-developed of all of these private sector products, and will likely become even more so with the recent purchase of Face.com. Facebook allows users to tag themselves in photos they upload and be tagged in others' photos. Facebook's "Tag Suggest" feature, introduced in December 2010, uses face recognition to automatically match uploaded photos to other photos a Facebook user is tagged in, grouping similar photos together and suggesting the name of a user's friend in the photo.[50] Facebook markets this tool by stating it will make sorting, tagging and finding photos easier,[51] but it does not make clear that the feature will create a unique biometric—a faceprint—for all its users.

Facebook has stirred up significant controversy with its face recognition tools, in large part because it turned these features on by default. It first enrolled all its users in the system without prior consent and then continued to opt-in users every time they uploaded a photograph. Users may opt-out of tagging on a photo-by-photo basis, but opting out of the system as a whole is complicated. Given the steps necessary to delete the face print "summary" data associated with each user's account[52] and the fact that Facebook uses persuasive language to try to dissuade users from deleting the data completely,[53] it is unlikely most users would go this far. And even if a user deletes the summary data, it is unclear whether taking this step will prevent Facebook from continuing to collect biometric data going forward.[54] As a result of these policies, Facebook has amassed possibly the largest database of face prints in the world[55]—with face prints for about 1/7 of the world population[56]—and will continue to collect more and more face prints every day as more users join the site.

Facebook and other companies using facial recognition combine this data with sensitive and personal biographic data and information on users' networks and associations, exacerbating privacy concerns. Facebook requires each of its users to sign up under their real names,[57] and then makes users' names, profile photos, gender and networks public by default.[58] Facebook is designed to promote social engagement, and as part of this users can and do provide extensive additional personal information, from email addresses and birthdates to partners' and family members' names, dating preferences, activities and interests, location information, and political and religious beliefs. Facebook also encourages users to communicate with each other through status updates, "likes," posts on other users' walls, and direct messages. Facebook then records all of this information as part of the user's profile, along with other less evident information, such as when users look at another person's profile; when they search for their friends; location, time and date information recorded in their photos; GPS information; and which device or computer they use to log into their account.[59] Through all of this, Facebook establishes associations between and among users and between users and the companies, organizations and causes they find relevant to their lives. All of this information is stored indefinitely by Facebook and, depending on a user's privacy settings, may be available beyond a user's friends or networks—even available to the public at large.

Some or all of this information may be shared with third parties such as other companies, app developers, and advertisers, depending on a user's privacy settings. In addition, the government regularly reviews and requests this data to verify citizenship applications,[60] for evidence in criminal cases,[61] and to look for threats to U.S. safety and security.[62]

As discussed in further detail below, few laws regulate private biometric collection on this scale. In general the public must rely on a company's privacy policies, terms of use, and user-managed privacy settings. However, as the public has seen with the extensive changes Facebook has made to its privacy settings and policies,[63] the fact that it implemented an opt-out based facial recognition system with little fanfare or explanation, and that the first facial recognition app developed to make tagging even easier (KLIK) was quickly hacked to allow access to private information in users' Facebook and Twitter accounts and automatically "recognize" anyone walking down the street,[64] industry self-regulation and consumer control are not enough to protect against critical privacy and security risks inherent in facial recognition data collection.

CONCERNS ABOUT BIOMETRICS, DATABASES, AND DATA SHARING

The extensive collection and sharing of biometric data at the local, national, and international level should raise significant concerns among Americans. Data accumulation and sharing can be good for solving crimes across jurisdictions or borders, but can also perpetuate racial and ethnic profiling, social stigma, and inaccuracies throughout all systems and can allow for government tracking and surveillance on a level not before possible.

Some of these concerns are endemic to all data collection and are merely exacerbated by combining biographic data with any non-changeable biometric. For example, courts have recognized the "social stigma" involved with merely having a record in a criminal database.[65] Additionally, data inaccuracies—such those common in immigration[66] and Automated Targeting System[67] records—become much more damaging and difficult to correct as they are perpetuated through cross-database sharing.

Data sharing can also mean that data collected for non-criminal purposes, such as immigration-related records or employment verification, are combined with and used for criminal or national-security purposes with little or no standards, oversight, or transparency. When some of this data comes from sources such as local fusion centers and private security guards in the form of Suspicious Activity Reports (SARs),[68] it can perpetuate racially or politically motivated targeting.[69]

Standardization of biometrics data (necessary to enable data sharing) causes additional concerns. Once data are standardized, they become much easier to use as linking identifiers, not just in interactions with the government but also across disparate databases and throughout society. For example, Social Security numbers were created to serve one purpose—to track wages for Social Security benefits—but are now used to identify a person for credit and background checks, insurance, to obtain food stamps and student loans, and for many other private and government purposes.[70] If biometrics become similarly standardized, they could replace Social Security numbers, and the next time someone applies for insurance, sees her doctor, or fills out an apartment rental application, she could be asked for her face print. This is problematic if records are ever compromised because, unlike a Social Security Number or other unique identifying number, a person cannot change her biometric data.[71] And the many recent security breaches and reports of falsified data show that the government and private sector can never fully protect against these kinds of data losses.[72] Data standardization also increases the ability of government and the private sector to locate and track a given person throughout her life.

And finally, extensive data retention periods[73] can lead to further problems; data that may be less identifying today, such as a photograph of a large crowd or political protest, could become more identifiable in the future as technology improves.

However, advanced biometrics like face recognition create additional concerns because the data may be collected in public without a person's knowledge. For example, the addition of crowd and security camera photographs into NGI means that anyone could end up in the database—even if they're not involved in a crime—by just happening to be in the wrong place at the wrong time, by fitting a stereotype that some in society have decided is a threat, or by, for example, engaging in suspect activities such as political protest in areas rife with cameras.[74] Given the FBI's history of misuse of data gathered on people during former FBI director J. Edgar Hoover's tenure[75] and the years following September 11, 2001,[76]—data collection and misuse based on religious beliefs, race, ethnicity and political leanings—Americans have good reason to be concerned about expanding government biometrics databases to include face recognition technology.

Technical issues specific to facial recognition make its use worrisome for Americans. For example, facial recognition's accuracy is strongly dependent on consistent lighting conditions and angles of view.[77] It may be less accurate with certain ethnicities and with large age discrepancies (for example, if a person is compared against a photo taken of himself when he was ten years younger). These issues can lead to a high rate of false positives—when, for example, the system falsely identifies someone as the perpetrator of a crime or as having overstayed their visa. In a 2009 New York University report on facial recognition, the researchers noted that facial recognition "performs rather poorly in more complex attempts to identify individuals who do not voluntarily self-identify . . . Specifically, the "face in the crowd" scenario, in which a face is picked out from a crowd in an uncontrolled environment."[78] The researchers concluded the challenges in controlling face imaging conditions and the lack of variation in faces over large populations of people make it unlikely that an accurate face recognition system will become an "operational reality for the foreseeable future."[79]

Some have also suggested the false-positive risk inherent in large facial recognition databases could result in even greater racial profiling by disproportionately shifting the burden of identification onto certain ethnicities.[80] This can alter the traditional presumption of innocence in criminal cases by placing more of a burden on the defendant to show he is *not* who the system identifies him to be. And this is true even if a face recognition system such as NGI offers several results for a search instead of one, because each of the people identified could be brought in for questioning, even if he or she was not involved in the crime. In light of this, German Federal Data Protection Commissioner Peter Schaar has noted that false positives in facial recognition systems pose a large problem for democratic societies: "in the event of a genuine hunt, [they] render innocent people suspects for a time, create a need for justification on their part and make further checks by the authorities unavoidable."[81]

LEGAL PROTECTIONS FOR PRIVACY IN BIOMETRIC DATA

Face recognition implicates important Constitutional values, including privacy, free speech and association, and the right to be free from unlawful searches and seizures. If the government starts regularly collecting and indexing public photographs—or obtains similar data from private companies—this would have a chilling effect on Americans' willingness to engage in public debate and to associate with others who's values, religion or political views may be considered questionable.

And yet the fact that face images can be captured without a detention and in public, or may be uploaded voluntarily to a third party such as Facebook, or may be collected and stored by private security firms and data aggregators, presents significant challenges in applying Constitutional protections.

The Fourth Amendment

The Fourth Amendment's prohibition of unreasonable searches and seizures presents a baseline protection for governmental biometrics collection in the United States.[82]

Although there are significant exceptions to Fourth Amendment protections that may make it difficult to map to biometric collection such as facial recognition,[83] a recent Supreme Court case, *U.S. v. Jones*,[84] and a few other cases[85] show that courts are The two justices further noted "Montanans do not reasonably expect that concerned about mass collection of identifying information—even collection of information revealed to the public or a third party—and are trying to identify solutions.

Cases like *Jones* suggest support for the premise that although we may tacitly consent to someone noticing our face or our movements when we walk around in public, it is unreasonable to assume that consent extends to our data being collected and retained in a database, to be subject to repeated searches for the rest of our lives.

This is buttressed by important privacy research showing that even though people voluntarily share a significant amount of information about themselves with others online, they still consider much of this information to be private in that they don't expect it to be shared outside of the networks they designate.[86]

In *United States v. Jones*,[87] nine justices held that a GPS device planted on a car without a warrant and used to track a suspect's movements constantly for 28 days violated the Fourth Amendment. For five of the justices, a person's expectation of privacy in not having his movements tracked constantly—even in public—was an important factor in determining the outcome of the case.[88]

Justice Sotomayor would have gone even further, questioning the continued validity of the third-party doctrine (holding that people lack a reasonable expectation of privacy in data such as bank records that they share with a third-party such as the bank).[89] She also recognized that:

> [a]wareness that the Government may be watching chills associational and expressive freedoms. And the Government's unrestrained power to assemble data that reveal private aspects of identity is susceptible to abuse.[90]

She questioned whether "people reasonably expect that their movements will be recorded and aggregated in a manner that enables the Government to ascertain, more or less at will, their political and religious beliefs, sexual habits, and so on."[91]

The fact that several members of the Court were willing to reexamine the reasonable expectation of privacy test[92] in light of newly intrusive technology could prove important for future legal challenges to biometrics collection. And some of the questions posed by the justices, both during oral argument and in their various opinions, could be used as models for establishing greater protections for data like facial recognition that is both shared with a third party such as Facebook and gathered in public.[93]

Other Laws May Provide Only Limited Protection to Face Recognition Data Collected by Government and the Private Sector

Privacy Act

The federal Privacy Act[94] "regulates the collection, maintenance, use, and dissemination of information about individuals by federal agencies . . . [and] authorizes civil suits by individuals . . . whose Privacy Act rights are infringed."[95] Although it applies to "personally identified information" collected by the government and gives citizens a way of gaining access to records and requesting their amendment, it has significant exceptions that minimize its effectiveness in actually protecting Americans' privacy rights. For example, it does not offer a remedy for "constitutional claims arising from alleged wrongs covered by the Privacy Act."[96] And law enforcement exemptions that allow agencies to shield criminal justice records from Privacy Act protections[97] make it unlikely it would offer any meaningful protections against face recognition data collection.

Stored Communications Act

The Stored Communications Act,[98] a law passed in 1986, would likely apply to protect face recognition-ready photographs and underlying face print data because it addresses voluntary and compelled disclosure of "stored wire and electronic communications and transactional records" held by or in storage with third-party service providers like Facebook and Google.[99] However, because the definition of communications and content of communications was written to apply to more traditional oral or written communications,[100] it is unclear how the Act would map to the underlying face print data within a photograph, and whether the government would be required to obtain a warrant or some lesser legal process prior to requesting a copy of this data.[101]

FTC Act

The Federal Trade Commission Act[102] gives the FTC some power to investigate and seek relief for practices that are "unfair" and "deceptive."[103] A trade practice is unfair if it "causes or is likely to cause substantial injury to consumers which is not reasonably avoidable by consumers themselves and is not outweighed by countervailing benefits to consumers or competition."[104] A trade practice is "deceptive" if it involves a "material representation, omission or practice that is likely to mislead a consumer acting reasonably in the circumstances, to the consumer's detriment."[105]

The FTC has settled several actions related to privacy in social media or web search that could show how the FTC might address an action related to collection of face recognition data.[106] However, FTC actions are limited, and, unlike court-developed law, the standards for determining whether a trade practice is unfair or deceptive area hazy. In addition, the FTC has so far failed to address the Electronic Privacy Information Center's complaint related to Facebook's face recognition program, despite the fact that it was filed over a year ago.[107] Further, commentators and media regularly recognize that the lack of universal privacy laws in the United States and the limited powers allotted to the FTC to regulate privacy issues, mean that companies have little incentive to change their practices.[108]

State Laws[109]

Three states—Illinois, Texas and Washington—have so far implemented laws that expressly apply to biometrics collection. While these laws have some holes, some of their protections could be used as models for federal legislation.

Illinois's law[110] applies to private entities and requires them to notify an individual in writing and obtain a written release before collecting the individual's biometric information, including "face geometry." Entities must disclose "purpose and length of term for which [the] biometric information is being collected, stored, and used," and may further not disclose a collected biometric without the individual's consent, unless the disclosure is required by law. Because this is a state law, it only applies to transactions in Illinois. However, as a state populated with almost 13 million people, Illinois residents could use this law to enforce changes that would likely affect the rest of the country. The law creates private right of action in encourage residents to pursue their own remedies against violations of the law, but with no agency designated to enforce compliance, it does not appear that the law has had much effect so far.

Texas' law[111] similarly regulates collection and use of biometric data, including "face geometry" & prohibits the collection of an individual's biometric data for a commercial purpose without first informing that individual and obtaining her consent. The law does not permit transfers of biometric data for any purpose other than: (1) to identify a deceased or missing individual if that individual previously consented to such identification; (2) for a transaction upon an individual's request or authorization; or (3) to disclose the data pursuant to a state or federal statute or for a law enforcement purpose pursuant to a warrant. Similar to the Illinois law, it creates private right of action for enforcement. It also allows the state Attorney General to bring an action for damages. However, it doesn't appear the Attorney General or any private citizen has yet brought an enforcement action under the law, despite the fact that a base-level reading of the statute would suggest it applies to Facebook's opt-out system.

Washington has had a law regulating biometric drivers' licenses since 2004,[112] which was recently updated to apply to face recognition.[113] The changes, which go into effect this summer, limit the purposes for which face recognition may be used,[114] set standards for the accuracy of the system and security of the data,[115] provide for a notice requirement,[116] and clarify the legal process required for state and federal law enforcement to access the data.[117] The new version of the statute also includes a reporting requirement.[118] However, where the old version of the law created a voluntary biometrics system for licenses in Washington, the new version appears to remove this voluntariness language.

California may also be worth looking at when considering different protections for biometrics data, especially given how proposed biometrics bills have fared in the state legislature. California has no law specifically protecting biometrics but California's strong constitutional privacy rights,[119] which also apply against private companies, could offer some protections for abuse of those rights. Since 1998, the California legislature has introduced several bills that would directly regulate biometrics collection. However, due in part to industry pushback, none of these laws has moved out of the legislature. Most recently, Senate Bill 761, which would require a company that collects or uses "sensitive information," including biometric data, to allow users to opt-out of its collection, use, and storage, has faced stiff opposition from technology companies and their trade organizations.[120]

The lack of robust protections at the state level makes it even more important for the federal government to consider legislation to prevent improper biometrics collection and search.

Proposals for Change

The over-collection of biometrics has become a real concern, but there are still opportunities—both technological and legal—for change.

Given the current uncertainty of Fourth Amendment jurisprudence in the context of biometrics and the fact that biometrics capabilities are undergoing "dramatic technological change,"[121] legislative action could be a good solution to curb the over- collection and over-use of biometrics in society today and in the future. If so, the federal government's response to two seminal wiretapping cases in the late 60s could be used as a model.[122] In the wake of *Katz v. United States*[123] and *New York v. Berger,*[124] the federal government enacted the Wiretap Act,[125] which laid out specific rules that govern federal wiretapping, including the evidence necessary to obtain a wiretap order, limits on a wiretap's duration, reporting requirements, and a notice provision.[126] Since then, law enforcement's ability to wiretap a suspect's phone or electronic device has been governed primarily by statute rather than Constitutional case law. Congress could also look to the Video Privacy Protection Act (VPPA),[127] enacted in 1988, which prohibits the "wrongful disclosure of video tape rental or sale records" or "similar audio-visual materials," requires a warrant before a video service provider may disclose personally identifiable information to law enforcement, and includes a civil remedies enforcement provision.

If legislation or regulations are proposed in the biometrics context, the following principles should be considered to protect privacy and security. These principles are based in part on key provisions of the Wiretap Act and VPPA and in part on the Fair Information Practice Principles (FIPPs), an internationally recognized set of privacy protecting principles.[128]

Limit the Collection of Biometrics

The collection of biometrics should be limited to the minimum necessary to achieve the government's stated purpose. For example, collecting more than one biometric from a given person is unnecessary in many situations. Similarly, the government's acquisition of biometrics from sources other than the individual to populate a database should be limited. For example, the government should not obtain biometrics en masse to populate its criminal databases from sources such as state DMV records, where the biometric was originally acquired for a non-criminal purpose, or from crowd photos or data collected by the private sector. Techniques should also be employed to avoid over-collection of face prints (such as from security cameras or crowd photos) by, for example, scrubbing the images of faces that are not central to an investigation.

Define Clear Rules on the Legal Process Required for Collection

Each type of biometric should be subject to clear rules on when it may be collected and which specific legal process—such as a warrant based on probable cause—is required prior to collection. Collection and retention should be specifically disallowed without legal process unless the collection falls under a few very limited and defined exceptions. For example, clear rules should be defined to govern when law enforcement or similar agencies may collect biometrics revealed to the public, such as a face print.

Limit the Amount and Type of Data Stored and Retained

For biometrics such as a face print that can reveal much more information about a person than his or her identity, rules should be set to limit the amount of data stored. Retention periods should be defined by statute and should be limited to no longer than necessary to achieve the goals of the program. Data that is deemed to be "safe" from a privacy perspective today could become highly identifying tomorrow. For example, a data set that includes crowd images could become much more identifying as technology improves. Similarly, data that was separate and siloed or unjoinable today might be easily joinable tomorrow. For this reason retention should be limited, and there should be clear and simple methods for a person to request removal of his or her biometric from the system if, for example, the person has been acquitted or is no longer under investigation.[129]

Limit the Combination of More than One Biometric in a Single Database

Different biometric data sources should be stored in separate databases. If biometrics need to be combined, that should happen on an ephemeral basis for a particular investigation. Similarly, biometric data should not be stored together with non-biometric contextual data that would increase the scope of a privacy invasion or the harm that would result if a data breach occurred. For example, combining facial recognition technology from public cameras with license plate information increases the potential for tracking and surveillance. This should be avoided or limited to specific individual investigations.

Define Clear Rules for Use and Sharing

Biometrics collected for one purpose should not be used for another purpose. For example, face prints collected for use in a criminal context should not automatically be used or shared with an agency to identify a person in an immigration context. Similarly, photos taken in a non-criminal context, such as for a driver's license, should not be shared with law enforcement without proper legal process. For private sector databases, users should be required to consent or opt-in to any face recognition system.

Enact Robust Security Procedures to Avoid Data Compromise

Because biometrics are immutable, data compromise is especially problematic. Using traditional security procedures, such as basic access controls that require strong passwords and exclude unauthorized users, as well as encrypting data transmitted throughout the system, is paramount. However security procedures specific to biometrics should also be enacted to protect the data. For example, data should be anonymized or stored separate from personal biographical information. Strategies should also be employed at the outset to counter data compromise after the fact and to prevent digital copies of biometrics. Biometric encryption[130] or "hashing" protocols that introduce controllable distortions into the biometric before matching can reduce the risk of problems later. The distortion parameters can easily be changed to make it technically difficult to recover the original privacy-sensitive data from the distorted data, should the data ever be breached or compromised.[131]

Mandate Notice Procedures

Because of the real risk that face prints will be collected without their knowledge, rules should define clear notice requirements to alert people to the fact that a face print has been collected. The notice provision should also make clear how long the biometric will be stored and how to request its removal from the database.

Define and Standardize Audit Trails and Accountability Throughout the System

All database transactions, including biometric input, access to and searches of the system, data transmission, etc. should be logged and recorded in a way that assures accountability. Privacy and security impact assessments, including independent certification of device design and accuracy, should be conducted regularly.

Ensure Independent Oversight

Government entities that collect or use biometrics must be subject to meaningful oversight from an independent entity. Individuals whose biometrics are compromised, whether by the government or the private sector should have a strong and meaningful private right of action.

CONCLUSION

Face recognition and its accompanying privacy concerns are not going away. Given this, it is imperative that government act now to limit unnecessary biometrics collection; instill proper protections on data collection, transfer, and search; ensure accountability; mandate independent oversight; require appropriate legal process before government collection; and define clear rules for data sharing at all levels. This is important to preserve the democratic and constitutional values that are bedrock to American society.

Thank you once again for the invitation to testify today. I am happy to respond to your questions.

End Notes

[1] FBI, *Integrated Automated Fingerprint Identification System*, http://www.fbi.gov/about-us/cjis/fingerprints_biometrics/iafis/iafis (last visited July 16, 2012); Elizabeth

Montalbano, "DHS Expands US-VISIT Biometric Capabilities," *Information Week* (Dec. 22, 2011), http://www.informationweek.com/news/government/ security/ 232300 942.

2 *See* Aliya Sternstein, "FBI to Launch Nationwide Facial Recognition Service," *Nextgov.com* (Oct. 7, 2011), *available at* http://www.nextgov.com/nextgov/ng_2011 1007_6100.php.

3 *See* Emily Steel, "How a New Police Tool for Face Recognition Works," *Wall St. J. Digits Blog* (July 13, 2011) http://blogs.wsj.com/digits/2011/07/13/how-a-new-police-tool-for-face-recognition-works/ (describing the Mobile Offender Recognition and Information System (MORIS), which attaches to an iPhone and allows face, fingerprint and iris scanning and identification). As I have written, law enforcement appears to be using these devices with little or no precursor level of suspicion. *See* Jennifer Lynch, *From Fingerprints to DNA: Biometric Data Collection in U.S. Immigrant Communities and Beyond*, Electronic Frontier Foundation & Immigration Policy Council Whitepaper, 3 (May 23, 2012) available at https://www.eff.org/wp/fingerprints-dna-biometric-data-collection-us-immigrant-communities- and-beyond.

4 Facebook, *Key Facts: Statistics* (last visited July 9, 2012) http://newsroom.fb.com/content/ default.aspx? NewsAreaId=22.

5 See Yaniv Taigman and Lior Wolf, "Leveraging Billions of Faces to Overcome Performance Barriers in Unconstrained Face Recognition," Face.com, http://face.com/research/faceR2011b.html (last visited Mar. 15, 2012).

6 Demian Bulwa, "Fingerprint check-in tried at 24 Hour Fitness," *S.F. Chron.* (Aug 23, 2010) http://www.sfgate.com/bayarea/article/Fingerprint-check-in-tried-at-24-Hour-Fit ness-3255272.php

7 For example, many Facebook users regularly upload photographs of their non-Facebook using babies and children and identify these images with a name in the description field for the photo. Others create Facebook profiles for their unborn children. *See* Steven Leckart, "The Facebook-Free Baby," *Wall St. J.*, http://online.wsj.com/article/ SB10001424052702304451104577392041180138910.html

8 This is a conservative estimate based on the latest U.S. population figures. It doesn't account for the fact that Facebook, which uses face recognition to scan all photographs uploaded, may be creating a face print for non-Facebook users as well.

9 18 U. S. C. §§2510–2522.

10 18 U.S.C. § 2710.

11 Elizabeth Montalbano, "DHS Expands US-VISIT Biometric Capabilities," *Information Week* (Dec. 22, 2011), http://www.informationweek.com/news/government/ security/ 232300942.

12 *Privacy Impact Assessment (PIA) for the Next Generation Identification (NGI) Interstate Photo System (IPS)* (hereinafter "2008 IPS PIA"), FBI (June 9, 2008), http://www.fbi.gov/foia/privacy-impact- assessments/interstate-photo-system.

13 *See* http://www.fbi.gov/about-us/cjis/fingerprints_biometrics/iafis/iafis (last visited Apr. 26, 2012).

14 *See* DHS, "Privacy Impact Assessment for the Automated Biometric Identification System (IDENT)," (July 31, 2006), *available at* http://www.dhs.gov/xlibrary/assets/privacy/ privacy_pia_usvisit_ident_final.pdf.

15 Customs and Border Protection (CBP) feeds biometrics data into IDENT while also maintaining its own database, called TECS, which includes personally identifiable information on and biometrics obtained from travelers crossing the border into the

United States. *See* DHS, "Privacy Impact Assessment for the TECS System: CBP Primary and Secondary Processing" ("TECS PIA"), DHS/CBP/PIA-009(a), (Dec. 22, 2010), *available at* http://www.dhs.gov/xlibrary/assets/privacy/privacy-pia-cbp-tecs-sar-update.pdf.

[16] DHS, "Privacy Impact Assessment for the Refugees, Asylum, and Parole System and the Asylum Pre- Screening System" (Nov. 24, 2009), *available at* http://www.dhs.gov/ xlibrary/assets/ privacy/privacy_pia_cis_rapsapss.pdf. USCIS also maintains its own database of "biometric images," including a digital photograph and signature, both of which appear on an applicant's naturalization certificates. *See* DHS, "Privacy Impact Assessment Update for the Computer Linked Application Information Management System, DHS/USCIS/PIA-015(a)" (Aug. 31, 2011), *available at* http://www.dhs.gov/ xlibrary/assets/privacy/privacy_pia_uscis_claimsupdate.pdf (describing capturing of "digitized biometric images" through the Benefits Biometric Support System (BBSS)).

[17] *See* DHS, "Government Agencies Using US-VISIT," http://www.dhs.gov/files/programs/ gc_1214422497220.shtm.

[18] Elizabeth Montalbano, "DHS Expands US-VISIT Biometric Capabilities," *Information Week* (Dec. 22, 2011), http://www.informationweek.com/news/government/ security/ 232300942.

[19] Existing biometric databases have allowed users to input some photographs, but they have been limited to traditional mug shots and have not incorporated facial recognition capabilities. *See* 2008 IPS PIA, http://www.fbi.gov/foia/privacy-impact-assessments/ interstate-photo-system.

[20] *Next Generation Identification*, FBI, http://www.fbi.gov/about-us/cjis/ fingerprints_ biometrics/ngi (last visited April 27, 2012).

[21] *See Statement: Interoperability Initiatives Unit* (December 2010), Bates No. SC-FBI-FPL-1043, available at http://uncoverthetruth.org/wp-content/uploads/S-Comm-Hot-Docs-Released-11-10-11.zip (download archive; unzip; open "SC-FBI-FPL-1043.pdf")

[22] *Id.*

[23] Once NGI is complete, it will include iris scans, palm prints, and voice data, in addition to fingerprints.

[24] According to *Nextgov.com*, these states include Michigan, Washington, Florida, and North Carolina. *See* Aliya Sternstein, "FBI to Launch Nationwide Facial Recognition Service," *Nextgov.com* (Oct. 7, 2011), *available at* http://www.nextgov.com/ nextgov/ng_20111007_6100.php. However, recently-released records from an FBI Criminal Justice Information Services Advisory Board meeting show that the FBI signed MOUs in December 2011 with Maryland, Michigan and Hawaii and may also be working with Oregon. *See* Jennifer Lynch, "FBI's Facial Recognition is Coming to a State Near You," *EFF.org* (forthcoming) https://www.eff.org/deeplinks/2012/07/ fbis_facial_recognition_coming_state_near_you.

[25] *See* 2008 IPS PIA, http://www.fbi.gov/foia/privacy-impact-assessments/interstate-photo-system.

[26] *See, e.g.*, Richard W. Vorder Bruegge, *Facial Recognition and Identification Initiatives*, 5, FBI available at http://biometrics.org/bc2010/presentations/DOJ/vorder_bruegge-Facial-Recognition-and-Identification- Initiatives.pdf (noting a goal of NGI is to "Identify[ ] subjects in public datasets").

[27] *See* 2008 IPS PIA, http://www.fbi.gov/foia/privacy-impact-assessments/interstate-photo-system.

[28] *See* Vorder Bruegge, *Facial Recognition and Identification Initiatives*, 5.

[29]*See* 2008 IPS PIA, http://www.fbi.gov/foia/privacy-impact-assessments/interstate-photo-system. The FBI has recognized that "electronic bulk searching of civil file images (such as via facial recognition technology) would constitute a significant new privacy consideration," *id.*, but the FBI has not yet released a new PIA.

[30] *See* "Accenture Awarded Biometric Identity System Contract from U.S. Department of Homeland Security," *Wall Street Journal Market Watch* (Dec. 21, 2011), at http://www.marketwatch.com/story/accenture-awarded-biometric-identity-system-contract- from-us-department-of-homeland-security-2011-12-21; Elizabeth Montalbano, "DHS Expands US-VISIT Biometric Capabilities," *Information Week* (Dec. 22, 2011), http://www.informationweek.com/news/government/security/232300942.

[31] William Ford, *State of Research, Development and Evaluation at NIJ*, 17, National Institute of Justice, http://biometrics.org/bc2010/presentations/DOJ/ford-State-of-Research-Development-and-Evaluation-at- NIJ.pdf.

[32] *See* Emily Steel, "How a New Police Tool for Face Recognition Works," *Wall St. J. Digits Blog* (July 13, 2011) http://blogs.wsj.com/digits/2011/07/13/how-a-new-police-tool-for-face-recognition-works/.

[33] James Fallows, "Technology Is Our Friend ... Except When It Isn't," *The Atlantic* (Aug. 27, 2011) http://www.theatlantic.com/technology/archive/2011/08/technology-is-our-friend-except-when-it- isnt/244233/.

[34] *Pinellas County Sheriff's Office: DHS Future Opportunities*, 10 (2010) http://biometrics. org/bc2010/presentations/DHS/mccallum-DHS-Future-Opportunities.pdf.

[35] Bob Sullivan, "Gov't cameras in your car? E-toll patent hints at Big Brotherish future," *MSN* (Oct. 14, 2011) http://redtape.msnbc.msn.com/_news/2011/10/14/8308841-govt-cameras-in-your-car-e-toll-patent- hints-at-big-brotherish-future.

[36] This was achieved through provisions in the USA PATRIOT Act, Pub. L. No. 107-56, 115 Stat. 272 (2001), several Executive Orders (Exec. Order No. 13356, 69 C.F.R. 53599 (2004), Exec. Order No. 13355, 69 C.F.R. 53593 (2004), Exec. Order No. 13354, 69 C.F.R. 53589 (2004), Exec. Order No. 13311, 68 C.F.R. 45149 (2003)), and the Intelligence Reform and Terrorism Prevention Act (IRTPA) of 2004, Pub. L. No. 108-458, 118 Stat. 3638 (2004).

[37] The National Institute for Standards and Technology (NIST), along with other standards setting bodies, has developed standards for the exchange of biometric data. See National Institute for Standards and Technology, ANSI/NIST-ITL 1-2011, *American National Standard for Information Systems: Data Format for the Interchange of Fingerprint, Facial & Other Biometric Information* (2011), available at http://www.nist.gov/customcf/get_pdf.cfm?pub_id=910136.

[38] For more on Secure Communities, *see* Michele Waslin, *The Secure Communities Program: Unanswered Questions and Continuing Concerns*, Immigration Policy Center (Nov. 2011). Similarly, DHS is now sharing its data on asylum applicants more broadly with non-DHS agencies, per federal regulation 8 CFR §208.6(a). According to a June 30, 2011, Privacy Impact Assessment, DHS now shares the entire Refugees, Asylum and Parole Services (RAPS) database with the National Counter Terrorism Center (NCTC), a division of the Office of the Director of National Intelligence, under a Memorandum of Understanding (MOU). Dep't of Homeland Sec., *Privacy Impact Assessment Update for the Refugees, Asylum, and Parole System and the Asylum Pre-Screening System,* DHS/USCIS/PIA-027(a) (June 30, 2011), available at http://www.dhs.gov/xlibrary/assets/privacy/privacy_pia_ raps_update_nctc.pdf. DHS has been sharing asylum data with the FBI since October 8, 2001, per an MOU signed

by the agencies on that date. See USCIS Asylum Division, *Fact Sheet on Confidentiality*, 6 (June 15, 2005), available at http://www.usa- federal-forms.com/uscis-index-html/uscis-fact-sheet-on-confidentiality-uscis-5413.html.

[39] Thomas Frank, "Four states adopt 'no-smiles' policy for driver's licenses," *USA Today*, (May 26, 2009) http://www.usatoday.com/news/nation/2009-05-25-licenses_N.htm.

[40] Mike Baker, "FBI uses facial-recognition technology on DMV photos," *USA Today* (Oct. 13, 2009), http://www.usatoday.com/tech/news/2009-10-13-fbi-dmv-facial-recog nition_N.htm. British Columbia attempted to use its DMV's face recognition database to identify people involved in the 2011 Stanley Cup riots, though this was later determined by B.C.'s Privacy Commissioner, Elizabeth Denham, to be a violation of Canada's privacy law. Jonathan Fowlie, "Court order required to use facial recognition to identify Stanley Cup rioters," *Vancouver Sun* (Feb. 17, 2012). http://www.vancouversun.com/business/Court+order+required+facial+recognition+identify+Stanley+rioters/6163995/story.html.

[41] *See* Kimberley Del Greco, "FBI Facial Services Program," FBI 5 (Sept. 29, 2011) *available at* http://www.biometrics.org/bc2011/presentations/DOJ/0929_ 1105_BrA_DelGreco.pdf.

[42] The FBI's Criminal Justice Information Service (CJIS) division has information-sharing relationships with 77 countries, and is working with several countries to allow real-time access to their respective biometrics databases. *See* FBI/CJIS Advisory Policy Board Identification Services Subcommittee, *Issue Paper: Biometrics Information Sharing Update* (Spring 2011), Bates No. SC-FBI-FPL-1088-89, available at http://uncover thetruth.org/wp-content/uploads/S-Comm-Hot-Docs-Released-11-10-11.zip (download archive; unzip; open "SC-FBI-FPL-1081.pdf") (noting these relationships are "in the form of both informal (ad hoc, verbal) agreements and formal agreements (Memoranda of Agreement, Memoranda of Understanding, Letter of Cooperation).").

[43] *Id.* at SC-FBI-FPL-1089; DHS, "Privacy Impact Assessment for the Automated Biometric Identification System (IDENT)," 8 (July 31, 2006) available at http://www.dhs.gov/xlibrary/assets/privacy/privacy_pia_ usvisit_ident_final.pdf. This kind of biometrics sharing could prove disastrous for repatriated refugees or immigrants from countries with a history of ethnic cleansing.

[44] My testimony focuses on face recognition, rather than face detection. However, for an excellent discussion of face detection and digital signage, *see* Pam Dixon, "The One-Way-Mirror Society: Privacy Implications of the new Digital Signage Networks," *World Privacy Forum* (Jan 27, 2010) *available at*: www.ftc.gov/os/comments/privacyroundtable/544506-00112.pdf; *see also* Harley Geiger, "Seeing is ID'ing: Facial Recognition & Privacy," Center for Democracy & Technology (Decl 6, 2011) https://www.cdt.org/report/seeing-iding-facial-recognition-and-privacy.

[45] Facebook, "Newsroom: Key Facts: Statistics," http://newsroom.fb.com/content/default.aspx? NewsAreaId=22 (last visited July 10, 2012).

[46] The Google+ "Find My Face" feature is different from Facebook's facial recognition tools because, unlike Facebook, users must first opt-in to the system. Chester Wisniewski, "Facial recognition comes to Google+, but unlike Facebook it's opt-in," *Naked Security* (Dec. 9 2011) http://nakedsecurity.sophos.com/2011/12/09/google-introduces-facial-recognition-feature-opt-in-unlike-facebooks-effort/; Matt Steiner, "Making photo tagging easier with Find My Face," *Google*, https://plus.google.com/11026004 3240685719403/posts/jKQ35ajJ4EU.

[47] *See* Matt Hickey, "Picasa Refresh Brings Facial Recognition," *TechCrunch* (Sept. 2, 2008) http://techcrunch.com/2008/09/02/picasa-refresh-brings-facial-recognition/; Wilson Rothman, "What To Know About iPhoto '09 Face Detection and Recognition," *Gizmodo* (Jan. 29, 2009) http://gizmodo.com/5141741/what-to-know-about-iphoto-09-face-detection-and-recognition.

[48] Christina Bonnington, "FaceVault App Brings Facial Recognition Unlocking to iOS," *Wired Gadget Lab Blog* (April 25, 2012) http://www.wired.com/gadgetlab/2012/04/facevault-app-face-recognition/.

[49] Face.com developed an app called KLIK that allowed users to tag people in photographs before the photo was even taken. However, after Facebook bought Face.com, the app was removed from the Apple app store. *See* David Murphy, "Facebook Kills Face.com Face-Recognition APIs, KLIK app," *PC Magazine* (July 7, 2012) http://www.pcmag.com/article2/0,2817,2406822,00.asp.

[50] Justin Mitchell, "Making Photo Tagging Easier," *The Facebook Blog* (Dec. 15, 2010), https://www.facebook.com/blog.php?post=467145887130.

[51] *Id.* (noting, "[n]ow if you upload pictures from your cousin's wedding, we'll group together pictures of the bride and suggest her name. Instead of typing her name 64 times, all you'll need to do is click 'Save' to tag all of your cousin's pictures at once.").

[52] *Id.* (noting users may turn off automatic tagging and remove tags added by others); *See also* Eva Galperin, "How to Disable Facebook's Facial Recognition Feature," *EFF* (June 9, 2011) www.eff.org/deeplinks/2011/06/how-disable-facebooks-facial-recognition-feature; Electronic Privacy Information Center (EPIC), "Complaint: In the Matter of Facebook, Inc, and the Facial Identification of Users," 12-15 (June 10, 2011) *available at* http://epic.org/privacy/facebook/EPIC_FB_FR_FTC_ Complaint_ 06_10_ 11.pdf.

[53] Facebook, "How Can I Turn Off Tag Suggestions?" https://www.facebook.com/help/?faq=187272841323203#How-can-I-turn-off-tag-suggestions ("Before you opt out of using this feature, we encourage you to consider how tag suggestions benefit you and your friends. Our tagging tools . . . are meant to make it easier for you to share your memories and experiences with your friends.")

[54] *Id.* EPIC Complaint at 16.

[55] Facebook users uploaded more than 300 million photos *every day* in the three months ending on March 31, 2012. Facebook, *Key Facts: Statistics* (last visited July 9, 2012) http://newsroom.fb.com/content/default.aspx?NewsAreaId=22. *See also* Facebook Photo Trends [INFOGRAPHIC], *PIXABLE* (Feb. 14, 2011) http://blog.pixable.com/2011/02/14/facebook-photo-trends-infographic/ (estimating that as of Summer 2011, users would have uploaded 100 billion photos to Facebook). Face.com, which developed face recognition tools for Facebook and was recently acquired the company, stated in March that it had indexed 31 billion face images. *See* Yaniv Taigman and Lior Wolf, "Leveraging Billions of Faces to Overcome Performance Barriers in Unconstrained Face Recognition," *Face.com*, http://face.com/research/faceR2011b.html (last visited Mar. 15, 2012).

[56] Facebook estimates it has 900 million users. Facebook, "Newsroom: Key Facts: Statistics," http://newsroom.fb.com/content/default.aspx? NewsAreaId=22 (last visited July 10, 2012). The world population is currently estimated at between six and seven billion people.

[57] *See* Emil Protalinski, "Facebook has over 845 million users," *ZDNet* (Feb. 1, 2012), http://www.zdnet.com/blog/facebook/facebook-has-over-845-million-users/8332;

Facebook "Statement of Rights and Responsibilities" (April 26, 2011), https://www.facebook.com/legal/terms ("Facebook users provide their real names and information . . . You will not provide any false personal information on Facebook[.]").

58 Facebook, "Understand Your Internet Search Listing: Is my information visible to people who aren't logged into Facebook?" https://www.facebook.com/help/privacy/public-search-listings (last visited July 10, 2012).

59 Facebook, "Information we receive and how it is used: Other information we receive about you," https://www.facebook.com/about/privacy/your-info#inforeceived (last visited July 10, 2012).

60 *See* Jennifer Lynch, "Applying for Citizenship? U.S. Citizenship and Immigration Wants to Be Your 'Friend,'"*EFF* (Oct. 12, 2010), https://www.eff.org/deeplinks/2010/10/applying-citizenship-u-s- citizenship-and (describing how USCIS agents "friend" applicants for citizenship on social networking sites in order to monitor them).

61 In warrant for Facebook data, the Department of Justice Criminal Division regularly requests all photos in which a user is tagged. *See* Jennifer Lynch, "DOJ Wants to Know Who's Rejecting Your Friend Requests," *EFF* (Jan. 24, 2012), https://www.eff.org/deeplinks/2012/01/doj-wants-know- who%E2%80%99s-rejecting-your-friend-requests.

62 Jennifer Lynch, "New FOIA Documents Reveal DHS Social Media Monitoring During Obama Inauguration," *EFF* (Oct. 13, 2010), https://www.eff.org/deeplinks/2010/10/new-foia-documents-reveal- dhs-social-media; Jennifer Lynch, "Government Uses Social Networking Sites for More than Investigations," EFF.org (Aug. 16, 2010), https://www.eff.org/deeplinks/2010/08/government-monitors-much-more-social-networks. The FBI is currently looking for software to make its mining of social-media data more efficient and to allow it to map communities of interest. *See* Jim Giles, "FBI releases plans to monitor social networks," *New Scientist* (Jan. 25, 2012), http://www.newscientist.com/blogs/onepercent/2012/01/fbi-releases-plans-to-monitor.html.

63 *See* Matt McKeon, Infographic: "The Evolution of Privacy on Facebook," http://mattmckeon.com/ facebook-privacy/ (last visited July 11, 2012).

64 *See* http://ashkansoltani.org/docs/face_palm.html (describing how independent privacy and security researcher Ashkan Soltani hacked Face.com's KLIK app); *See also* Alessandro Acquisti, Face Recognition Study—FAQ, http://www.heinz.cmu.edu/~acquisti/face-recognition-study-FAQ/; Will Oremus, "Facebooked in the Crowd," *Slate.com* (June 19, 2012) http://www.slate.com/blogs/future_tense/ 2012/ 06/ 19/facebook_buys_face_com_will_mobile_facial_recognition_kill_privacy_.html (describing Acquisti's research combining "off-the-shelf facial recognition software . . . with Facebook data and a computer algorithm to guess, not only people's names, but in some cases their social security numbers, based solely on snapshots taken with a webcam").

65 *Menard v. Saxbe*, 498 F.2d 1017, 1024 (D.C. Cir. 1974) ("disabilities flowing from a record of arrest have been well documented: There is an undoubted 'social stigma' involved in an arrest record. It is common knowledge that a man with an arrest record is much more apt to be subject to police scrutiny -- the first to be questioned and the last to be eliminated as a subject in any investigation. . . . Most significant is its use in connection with subsequent inquiries on applications for employment and licenses to engage in certain fields of work. An arrest record often proves to be a substantial barrier to employment." Id. at 1024" (internal citations and footnotes omitted)).

[66] *See generally* Joan Friedland, National Immigration Law Center, *INS Data: The Track Record*, available at www.nilc.org/document.html?id=233 (citing multiple Government Accountability Office and Inspector General reports on inaccuracies in immigration records). These problems persist. *See generally, e.g,* U.S. Government Accountability Office (GAO), *Federal Agencies Have Taken Steps to Improve E-Verify, but Significant Challenges Remain*, GAO-11-146 (Jan. 18, 2011), available at http://www.gao.gov/ products/ GAO-11-146 (noting errors in USCIS's e-Verify system and difficulties in correcting those errors). This has happened with the Secure Communities program, where approximately 3,600 United States citizens have been caught up in the program due to incorrect immigration records. *See, e.g.,* Aarti Kohli, et al. *Secure Communities by the Numbers: An Analysis of Demographics and Due Process*, at p.4, Chief Justice Earl Warren Institute on Law and Social Policy, UC Berkeley School of Law (Oct. 2011), *available at* www.law.berkeley.edu/ files/Secure_ Communities_by_the_Numbers.pdf.

[67] The Automated Targeting System (ATS), which assigns everyone who crosses United States borders, a computer-generated 'risk assessment' score. Data collected by ATS is "stored for 15 years, even for individuals who have not been flagged as a threat or potential risk." *See* Shana Dines, "Interim Report on the Automated Targeting System: Documents Released through EFF's FOIA Efforts," *EFF.org* (Summer 2009), https://www.eff.org/pages/interim-report-autom. Under ATS, individuals have no way to access information about their "risk assessment" scores or to correct any false information about them. *See* "Lawsuit Demands Answers About Government's Secret 'Risk Assessment' Scores," *EFF* (Dec. 19, 2006), https://www.eff.org/press/ archives/2006/12/19.

[68] *See, e.g.,* G.W. Schulz & Andrew Becker, "Finding Meaning In Suspicious Activity Reports," *NPR* (Sept. 7, 2011), http://www.npr.org/2011/09/07/140237086/finding-meaning-in-suspicious-activity-reports; ACLU, *More About Suspicious Activity Reporting* (June 29, 2010), http://www.aclu.org/spy-files/more- about-suspicious-activity-reporting.

[69] *See, e.g.,* Robert Smith, "Julia Shearson tells how a weekend trip to Canada became 5-year fight for rights," *The Plain Dealer* (June 4, 2011), available at http://blog. cleveland.com/metro/2011/06/julia_shearson_tells_how_a_wee.html (describing how Executive Director of the Cleveland Council on American-Islamic Relations (CAIR) ended up on an FBI terrorist watchlist and her struggle to correct inaccuracies in her government files).

[70] *See* "Legal requirements to provide your SSN," *Social Security Online*, http://ssa-custhelp.ssa.gov/app/answers/detail/a_id/78/~/legal-requirements-to-provide-your-ssn.

[71] Records could be compromised in several ways. For example, faceprints are stored as algorithms rather than images. These algorithms could be changed within the database such that when a person tries to use her biometric to identify herself, the database doesn't recognize her or thinks she's someone else.

[72] *See, e.g.,* David Stout and Tom Zeller Jr., "Vast Data Cache About Veterans Is Stolen," *N.Y. Times* (May 23, 2006), available at https://www.nytimes.com/2006/05/ 23/washington/23identity.html; *see also* European Parliament News, *MEPs question Commission over problems with biometric passports* (Apr. 19, 2012) (noting that "In France 500,000 to 1 million of the 6.5 million biometric passports in circulation are estimated to be false, having been obtained on the basis of fraudulent documents.") *available at* http://www.europarl.europa.eu/news/en/headlines/content/ 20120413STO

42897/html/MEPs-question- Commission-over-problems-with-biometric-passports. *See also* discussion of KLIK app hack and Alessandro Acquisti's work supra n. 64.

[73] Biometric records stored in IDENT are retained for 75 years or until the statute of limitations for all criminal violations has expired. DHS, *Privacy Impact Assessment (PIA) for the Automated Biometric Identification System (IDENT)* (Jul. 31, 2006), http://www.dhs.gov/xlibrary/assets/privacy/ privacy_pia_usvisit_ident_final.pdf. Civil fingerprints stored in IAFIS are not destroyed until "the individual reaches 75 years of age," and criminal fingerprints are not destroyed until "the individual reaches 99 years of age." FBI, *Privacy Impact Assessment for the Fingerprint Identification Records System (FIRS) Integrated Automated Fingerprint Identification System (IAFIS) Outsourcing for Noncriminal Justice Purpose—Channeling* (May 5, 2008), http://www.fbi.gov/foia/privacy-impact- assessments/firs-iafis.

[74] For example, in Lower Manhattan, where the Occupy protests started, the New York Police Department has installed as many as 3,000 security cameras. *See* Noah Shachtman, "NYC Is Getting a New High-Tech Defense Perimeter. Let's Hope It Works," *Wired* (Apr. 21, 2008), http://www.wired.com/politics/security/magazine/16-05/ff_manhattansecurity.

[75] *See generally* Tim Weiner, *Enemies: A History of the FBI* (2012).

[76] *See, e.g.,* DOJ, Office of Inspector General (OIG), *A Review of the Federal Bureau of Investigation's Use of National Security Letters*, Special Report (March 2007); DOJ, OIG, *A Review of the FBI's Use of National Security Letters: Assessment of Corrective Actions and Examination of NSL Usage in 2006,* Special Report, (March 2008); DOJ, OIG, *A Review of the Federal Bureau of Investigation's Use of Exigent Letters and Other Informal Requests for Telephone Records* (January 2010).

[77] Face recognition technologies perform well when all the photographs are taken with similar lighting and shot from a frontal perspective (like a mug shot). However, with different lighting, shadows, different backgrounds, different poses or expressions, or as a person ages, the error rates are significant. See, e.g., P. Jonathon Phillips, et al., "An Introduction to the Good, the Bad, & the Ugly Face Recognition: Challenge Problem," *National Institute of Standards & Testing* (Dec. 2011), available at www.nist.gov/itl/iad/ig/upload/05771424.pdf *(*noting only 15% accuracy for face image pairs that are "difficult to match"). Security researcher Bruce Schneier has noted that even a 90% accurate system "will sound a million false alarms for every real terrorist" and that it is "unlikely that terrorists will pose for crisp, clear photos." Bruce Schneier, *Beyond Fear: Thinking Sensibly About Security in an Uncertain World,* 190 (2003).

[78] Lucas D. Introna & Helen Nissenbaum, *Facial Recognition Technology: A Survey of Policy and Implementation Issues*, p. 3, N.Y.U. (April 2009), available at http://www.nyu.edu/ccpr/pubs/Niss_04.08.09.pdf.

[79] *Id.* at 47. In layman's terms, this means that because so many people within a given population look alike, the probability that any facial recognition system will regularly misidentify people becomes much higher as the data set (the population of people you are checking against) gets larger.

[80] *Id.* at 45-46.

[81] *Id.* at 37.

[82] The Supreme Court has noted that the collection of biometrics like fingerprints has some Fourth Amendment protection, *see Davis v. Mississippi*, 394 U.S. 721, 723-24 (1969) (excluding from evidence fingerprints obtained during an illegal detention), however, the Court has declined to define the boundaries of that protection and suggested in dicta

that because "[f]ingerprinting involves none of the probing into an individual's private life and thoughts that marks an interrogation or search[,]" perhaps that protection is limited. *Id.* at 727. Courts have found greater protection in the collection of biological material that "can reveal a host of private medical facts about an [individual]," finding the collection "intrudes upon expectations of privacy that society has long recognized as reasonable." *Skinner v. Ry. Labor Executives' Ass'n*, 489 U.S. 602, 617 (1989).

[83] For example, in each of the key Supreme Court cases to address collection of biometrics or biological material, the legal analysis hinged in large part on the detention required to obtain the biometric data or on "a meaningful interference with [one's] possessory interest in his bodily fluids." *Skinner v. Ry. Labor Executives' Ass'n*, 489 U.S. 602, 618 n.4 (1989). However, biometrics such as face prints can be obtained without an initial detention and without the subject's knowledge while the subject is in a public place. Several cases have held that suspects have no legitimate expectation of privacy in biological material obtained under similar circumstances, *See* Erin Murphy, *The New Forensics: Criminal Justice, False Certainty, and the Second Generation of Scientific Evidence*, 95 Calif. L. Rev. 721, 736 n.63 and accompanying text (2007) (citing cases), or in discarded or abandoned material (such as garbage) or evidence in public view, making Fourth Amendment protection for face prints more tenuous. See, e.g., *California v. Greenwood*, 486 U.S. 35 (1988) (no reasonable expectation of privacy in garbage left on the street); *California v. Ciraolo*, 476 U.S. 207 (1986) (no expectation of privacy in backyard that can be viewed from a plane flying above); Elizabeth Joh, *Reclaiming "Abandoned" DNA: The Fourth Amendment and Genetic Privacy*, 100 Nw. U. L. Rev. 857, 863-64 (2006) (distinguishing cases where courts have found a "meaningful interference with an individual's possessory interests" from cases where "suspects 'knowingly expose' items to public view").

[84] 565 U.S. ____ (2012).

[85] *See, e.g.*, *United States v. Warshak*, 631 F.3d 266 (6th Cir. 2010) (holding email users have the same reasonable expectation of privacy in their stored email as they do in their phone calls and postal mail); *Montana State Fund v. Simms*, 270 P.3d 64 (Mont. 2012) (in concurring, two justices applied *US v. Jones*, finding the State Fund's "admitted practice of tracking, monitoring, and videotaping workers' compensation claimants as they go about their daily lives" implicated constitutional rights despite the fact that the videotaping occurred in public. The two justices further noted "Montanans do not reasonably expect that state government, in its unfettered discretion and without a warrant, is recording and aggregating their everyday activities and public movements in a manner which enables the State to ascertain and catalog their political and religious beliefs, their sexual habits, and other private aspects of identity." *Id.* at 71).

[86] danah boyd, *The Future of Privacy: How Privacy Norms Can Inform Regulation*, Oct. 29, 2010, available at http://www.danah.org/papers/talks/2010/PrivacyGenerations.html

[87] 565 U.S. ____ (2012).

[88] *Id.* (slip op. at 2-3) (Sotomayor, J. concurring); *Id.* (slip op. at 9–12) (Alito, J., concurring).

[89] *See also United States v. Warshak*, 631 F.3d 266 (6th Cir. 2010); *Montana State Fund v. Simms*, 270 P.3d 64 (Mont. 2012).

[90] *United States v. Jones,* 132 S. Ct. 945 (Sotomayor, J. concurring), 956; *see also NAACP v. Alabama*, 357 U.S. 449 (1958) (holding that requiring NAACP to disclose membership lists to the government would violate due process and a right to "associate freely with others").

[91] *Id.*

[92] See *Katz v. United States*, 389 U. S. 347, 361 (1967) (Harlan, J., concurring).

[93] Recently, privacy law scholars proposed several ways that Fourth Amendment doctrine could evolve in the wake of *Jones*. *See* www.usvjones.com. Susan Freiwald, who submitted the winning proposal, identified a four-factor test that incorporated factors the Supreme Court and appellate courts already identified. *See* Susan Freiwald, "The Four Factor Test," http://usvjones.com/2012/06/04/the-four-factor-test/ (noting that this four factor test "identifies when a surveillance method intrudes on Fourth Amendment rights and requires heightened judicial oversight to protect against abuse." These factors include whether the surveillance is *hidden* (the target is unaware of it), whether it is *intrusive* (offering access to "things people consider private"), *continuous*, and *indiscriminate* (gathering up "more information than necessary to establish guilt"). These factors could apply to restrict the collection of photographs taken from a hidden security camera that is always on and includes facial recognition.

[94] 5 U.S.C. §552a.

[95] *Jimenez v. Exec. Office for United States Attys.*, 764 F. Supp. 2d 174, 183 (D.D.C. 2011) (citing *Wilson v. Libby*, 535 F.3d 697, 707 (D.C. Cir. 2008)).

[96] *Id.* at 183.

[97] *See e.g.*, 28 C.F.R. § 16.81(a)(4) & (b)(3) (exempting from Privacy Act records maintained in US attorney criminal files).

[98] 18 U.S.C. §§ 2701–2712.

[99] *Id.* at §2703.

[100] *See* EFF, "Content of Communications," *EFF Internet Law Treatise,* https://ilt.eff.org/index.php/Privacy:_Data_Terminology# Content_of_Communications.

[101] For further discussion of the Stored Communications Act, *see* EFF, "Privacy: Stored Communications Act," *EFF Internet Law Treatise*, https://ilt.eff.org/index.php/ Privacy:_ Stored_Communications_Act.

[102] 15 U.S.C. §§ 41-58.

[103] *See* 15 U.S.C. § 45 (more commonly known as Section 5 of the FTCA) which declares "unfair or deceptive acts or practices in or affecting commerce" to be unlawful.

[104] 15 U.S.C. § 45(n).

[105] *See* Fed. Trade Comm'n, FTC Policy Statement on Deception, Letter from Fed. Trade Comm'n to Hon. John D. Dingell, Chairman, H. Comm. On Energy and Commerce (Oct. 14, 1983), http://www.ftc.gov/bcp/policystmt/addecept.htm ("Deception Statement").

[106] *See* Julianne Pepitone, "Facebook settles FTC charges over 2009 privacy breaches," *CNN.com* (Nov. 29, 2011) http://money.cnn.com/2011/11/29/technology/facebook_ settlement/index.htm; FTC, "FTC Gives Final Approval to Settlement with Google over Buzz Rollout" (Oct. 24, 2011) http://www.ftc.gov/opa/2011/10/buzz.shtm.

[107] *See* EPIC, "Complaint: In re Facebook and the Facial Identification of Users," (June 10, 2011) https://epic.org/privacy/facebook/facebook_and_facial_recognitio. html# comp laint.

[108] *See, e.g.*, Ryan Singel, "FTC's $22M Privacy Settlement With Google Is Just Puppet Waving," *Wired Threat Level Blog* (July 10, 2012) http://www.wired.com/threatlevel/ 2012/07/ftc-google-fine/ (noting that even the FTC's proposed $22.5 million fine to Google for violating the Google Buzz consent decree did not prevent the company from combining all user data).

[109] Special thanks to EFF Intern Yana Welinder for help with this section on state laws. *See* Yana Welinder, *A Face Tells More than a Thousand Posts: Developing Face Recognition Privacy in Social Networks*, (working paper) (July 16, 2012) *available at* http://papers.ssrn.com/sol3/papers.cfm?abstract_id=2109108.

[110] 740 Ill. Comp. Stat. 14/5.

[111] Tex. Bus. & Com. Code Ann. § 503.001.

[112] Chris Ingalls, "State shuts down successful crime-fighting tool," *King5.com* (Sept. 12, 2011) http://www.king5.com/news/investigators/Facial-recognition-program-shutdown--129663433.html.

[113] *See* Rev. Code. Wash. § 46.20.037 (revised by Substitute Senate Bill 6150, to take effect in 2012).

[114] *Id.* Sec. 1 ("Any facial recognition matching system selected by the department must be used only to verify the identity of an applicant for or holder of a driver's license to determine whether the person has been issued a driver's license, permit, or identicard under a different name or names.")

[115] *Id.* Sec. 2.

[116] *Id.* Sec. 3, 5 (notice "must address how the facial recognition matching system works, all ways in which the department may use results from the facial recognition matching system, how an investigation based on results from the facial recognition matching system would be conducted, and a person's right to appeal any determinations made under this chapter").

[117] *Id.* Sec. 4 (face recognition data "[m]ay only be disclosed [to state and local law enforcement] when authorized by a court order; [and m]ay only be disclosed to a federal government agency if specifically required under federal law").

[118] *Id.*

[119] Cal. Const. Art 1, sec. 1.

[120] *See* Opp'n Letter to Sen. Lowenthal (Apr. 27, 2011), *available at* http://static.arstechnica.com/oppositionletter.pdf.

[121] *Jones*, 565 U.S. , (slip op. at 13) (Alito, J., concurring).

[122] In Justice Alito's concurrence in *Jones*, he specifically referenced post-*Katz* wiretap laws and called out for legislative action, noting "[i]n circumstances involving dramatic technological change, the best solution to privacy concerns may be legislative." *Id.* (slip op. at 11, 13) (Alito, J., concurring).

[123] 389 U.S. 347 (1967).

[124] 388 U.S. 41 (1967). *Berger* was unique in that it struck down a state wiretapping law as facially unconstitutional. In striking down the law, the Court laid out specific principles that would make a future wiretapping statute constitutional under the Fourth Amendment.

[125] 18 U. S. C. §§2510–2522.

[126] *See, e.g.*, Orin S. Kerr, *The Fourth Amendment and New Technologies: Constitutional Myths and the Case for Caution*, 102 Mich. L. Rev. 801, 851-52 (2004).

[127] 18 U.S.C. § 2710.

[128] *See* Privacy Act of 1974, 5 U.S.C. § 552a (2010). *See also* Organization for Economic Co-operation and Development, *OECD Guidelines on the Protection of Privacy and Transborder Flows of Personal Data* (1980) available at http://www.oecd.org/document/18/0,3343,en_2649_34255_1815186_1_1_1_1,00.html. The full version of the FIPPs as used by DHS includes eight principles: Transparency, Individual Participation, Purpose Specification, Data Minimization, Use Limitation, Data Quality

and Integrity, Security, and Accountability and Auditing. *See* Hugo Teufel III, Chief Privacy Officer, DHS, Mem. No. 2008-01, Privacy Policy Guidance Memorandum (Dec. 29, 2008), available at http://www.dhs.gov/xlibrary/assets/privacy/privacy_policyguide_2008-01.pdf. *See also Fair Information Practice Principles*, FTC, http://www.ftc.gov/reports/privacy3/fairinfo.shtm (last modified June 25, 2007).

[129] For example, in *S. and Marper v. United Kingdom*, the European Court of Human Rights held that retaining cellular samples and DNA and fingerprint profiles of people acquitted or people who have had their charges dropped violated Article 8 of the European Convention on Human Rights. *S. and Marper. v. United Kingdom*, App. Nos. 30562/04 and 30566/04, 48 Eur. H.R. Rep. 50, 77, 86 (2009).

[130] *See, e.g.*, Information and Privacy Commissioner, Ontario, Canada, *Privacy-Protective Facial Recognition: Biometric Encryption—Proof of Concept* (Nov. 2010), available at www.ipc.on.ca/images/Resources/pbd-olg-facial-recog.pdf.

[131] *See, e.g.*, Center for Unified Biometrics and Sensors, "Cancellable Biometrics," SUNY Buffalo, http://www.cubs.buffalo.edu/cancellable.shtml (last visited Mar. 15, 2012).

In: Facial Recognition Technology ... ISBN: 978-1-62417-572-5
Editor: Meena N. Harnois © 2013 Nova Science Publishers, Inc.

Chapter 4

STATEMENT OF NITA A. FARAHANY, PROFESSOR OF LAW, DUKE LAW SCHOOL. HEARING ON "WHAT FACIAL RECOGNITION TECHNOLOGY MEANS FOR PRIVACY AND CIVIL LIBERTIES"*

Chairman Franken, Ranking Member Coburn, and distinguished members of the Subcommittee. Thank you for the opportunity to express my views about facial recognition technology, and its implications for privacy and civil liberties. My name is Nita Farahany. I am a Professor of Law and Research Professor of Genome Sciences and Policy at Duke University, and my research focuses on the ethical and legal implications of emerging technologies. I am also a member of the Presidential Commission for the Study of Bioethical Issues. I appear before you today in my individual, scholarly capacity.

I will focus my comments on the legal and constitutional implications of facial recognition technologies. I hope to show that, as a general matter, law enforcement use of these technologies is not, in itself, a Fourth Amendment search, let alone an unreasonable one. Although the Court has not yet addressed facial recognition technology, the doctrine regarding analogous identifying information and "open fields" supports this view.

* This is an edited, reformatted and augmented version of testimony given on July 18, 2012 before the Senate Judiciary Committee, Subcommittee on Privacy, Technology and the Law.

I. FACIAL RECOGNITION TECHNOLOGY AND BIOMETRIC IDENTIFYING INFORMATION

I will begin by explaining why I believe that law enforcement use of facial recognition technology is not a Fourth Amendment search, let alone an unreasonable one, and I will explain how Supreme Court doctrine is consistent with this view. Because the Fourth Amendment safeguards the right of the people to be secure against unreasonable searches and seizures by the government—but not by private or commercial actors—my remarks will focus on law enforcement use of this technology.

A. Facial Recognition Technology in Context: Identifying Information

Facial recognition technology uses software to try to match one's facial characteristics (such as the distance between eyes, the bridge of the nose, cheekbones, and other facial topography) with an existing database of facial data to identify an individual.[1]

As the technology has developed, the accuracy of identity matching has improved by capturing biometric features of faces such as skin texture, and by overcoming previous hurdles posed by dim lighting and subject movement with infrared imagery.[2]

Facial recognition technology is part of a broader class of identification technology, which uses the physical characteristics of an individual to match identity.[3] Other biometric identification technologies, for example, fingerprinting and iris scans, are already being used by law enforcement.[4] In fact, police forces across the country are rolling out new mobile investigative devices, some of which simply attach to the back of an iPhone, that can scan a fingerprint, an iris, or a face, and compare the results against existing databases.[5] These biometric devices and techniques add more precision to the vast array of identifying information investigators regularly obtain about individuals.[6]

A novel feature of facial recognition technology is that the first step of the process— scanning a face of interest—is usually done from a distance and without the awareness of the individual being scanned. The technology does not require physical contact, close physical proximity, or physical detention of an individual to scan their face. Infrared imagery even allows scanning to occur while a person is moving freely and is in dim lighting.

Facial recognition technology captures the sort of identifying information that is the bread and butter of law enforcement: information about the characteristics, physical likeness, and other descriptive features of a suspect.[7] It is routine practice for investigators to collect identifying information from individuals including their name, birth date, weight, height, clothing size, shoe size, blood type, and traces of shed DNA.[8] Whether collected through police-targeted or automatic photographing, facial recognition technology collects identifying evidence about individuals, a class of evidence that has traditionally received only minimal constitutional protection.

The second step of the process—comparing the scanned face to an existing database of photographs to find a match—is akin to the now-commonplace use by law enforcement of other identifying databases. Police routinely check local and national databases to find the identity of individuals by using their license plates, social security numbers, fingerprints, iris scans, and DNA to probe databases that contain such information. And all of this is nothing more than an automated version of what police have done for centuries: compare information acquired in the world with information held at police headquarters, looking for a match.

Neither the first step—scanning—nor the second step—querying government databases—implicates individual interests safeguarded by the Fourth Amendment. Neither step is properly characterized as a Fourth Amendment "search," let alone an "unreasonable" one, because under current law, neither step intrudes upon a legally cognizable privacy interest.

B. Scanning and Database Queries Are Not Fourth Amendment "Searches"

a. Scanning from Afar

The Fourth Amendment guarantees "[t]he right of the people to be secure in their persons, houses, papers, and effects, against unreasonable searches and seizures."[9] A Fourth Amendment search only occurs when the government intrudes upon a legally cognizable interest of an individual. Neither scanning an individual's face in public afar, nor querying a database to see if there is a match, intrudes on a cognizable privacy interest of an individual, so neither step constitutes a Fourth Amendment search.

If the police use facial recognition technology to scan an individual's face while in a public place, and that individual is not detained or touched as he is scanned, then no Fourth Amendment search has occurred. Neither his person nor his effects have been disturbed, and he lacks any legal source to support a reasonable expectation that his facial features will be hidden from government observation.[10] He has chosen to present his face to the world, and he must expect that the world, including the police, may be watching. Cameras and machines may now be doing the scanning, but for constitutional purposes, this is no different from an alert police officer "scanning" faces in a public place. This has never been thought to be a Fourth Amendment search.

By analogy, if the police observe an individual while he in public, and then return to the police station to flip through mug shot books to identify the individual they saw, no Fourth Amendment search or seizure has occurred. When the individual appeared in a public place, he relinquished his privacy interest in his seclusion. By subsequently observing the individual or comparing him against mug shot books, the police have not intruded on any legal interest retained by the individual. When an individual voluntarily forgoes seclusion, he cannot insist that the police avert their eyes.[11]

If the police were to use sense-enhanced technology to surreptitiously peer through the walls of a person's home, the police will intrude upon his real property interests and seclusion he has sought by taking refuge in his home.[12] Those interests support a reasonable expectation of privacy that an individual can exclude the police from surreptitiously observing him while secluded from public view inside his home. Because his reasonable expectation of privacy will be invaded if the police observe him this manner, a Fourth Amendment search will have occurred. Yet notice a crucial difference here between facial scanning in public and spying on an individual while inside their home: to obtain the facial scan, the police do not in any way invade the personal security or property of the individual.[13] They merely observe the individual from afar, in public, and capture the image that they see.[14] This does not constitute a search.

But even if it did constitute a search, it would likely be a constitutionally reasonable one, consistent with the Fourth Amendment. An unreasonable search occurs when the degree of insult or intrusion of an individual's legal interest outweighs the societal interest in the evidence being sought. So if the police surreptitiously observe an individual inside his home, the search will only be unreasonable if the degree of intrusion on the individual outweighs the societal need for the evidence sought. Since the Court primarily uses property rights to inform Fourth Amendment privacy interests, it likewise measures the intrusiveness of the search by assessing the physical intrusion upon the property, instead of the personal indignity that one may have endured by having their personal information revealed. Because protection of the home receives the most stringent Fourth Amendment protection, even a legitimate societal need for observing an individual inside their home is unlikely to be found reasonable.[15] By contrast, searching or seizing a person's likeness by scanning their face from afar does not interfere with their personal security or a right to exclude others they may otherwise enjoy. No physical violence or even physical interference occurs: Mere observation is not tantamount to a search, and certainly not an unreasonable one.

b. Probing a Database

Is using a captured image to query a photographic database for a match a Fourth Amendment search? Just as a criminal suspect lacks a cognizable legal basis to complain when the police peruse mug shots at the police station, neither can an individual successfully claim that a police query of a photographic database constitutes a search of *their* persons, houses, or effects.

If the information within the database and the probe used to search it have been lawfully collected, there is no additional Fourth Amendment interest that a claimant has in preventing the police from matching their identity.[16] Ultimately, the privacy concern usually advanced regarding the inclusion in or probing of forensic databases is whether an individual has a right to *secrecy* of identifying information. But the Court has never recognized a Fourth Amendment privacy interest in the secrecy of identifying information.[17] So the probe of a photographic database is not properly considered a separate Fourth Amendment search.

c. Scanning During "Stops"

The police might instead choose to use facial scanning technology during a brief investigative stop, which requires a slightly different constitutional analysis. Beginning with *Terry v. Ohio*,[18] the Court has held that if a police officer has a reasonable suspicion that someone has committed, is committing, or is about to commit a crime, the officer may detain the individual without a warrant for a brief investigative stop. While such stops are Fourth Amendment "searches" and a person is "seized" while they are detained, a warrantless stop based on reasonable suspicion may nevertheless be a reasonable search and/or seizure.

During a *Terry* stop, the police may require an individual to disclose his identity,[19] and a facial recognition scan for identification is not meaningfully different. Indeed, a suspect can be "arrested and prosecuted for refus[ing]" to disclose his identity during a stop based on reasonable suspicion.[20] As the Court has explained, states may require a suspect to disclose his name during an investigative stop because the individual privacy interest in identity is negligible compared to the legitimate government interests promoted by the inquiry.[21]

If a police officer detains an individual based on reasonable suspicion, using facial recognition technology during that stop would not meaningfully change the Fourth Amendment reasonableness of the search and seizure. The individual privacy interest that the Court recognizes during "stop and frisk" detentions is the personal security the individual and an interest against interference with his free movement, not the secrecy of their personal information or his personal identity. In other words, the Court has not included secrecy of personally identifying information as a relevant privacy interest in determining the reasonableness of a "stop and frisk" detention.[22] If the police can take the more intrusive step of requiring a suspect to state his name, it can surely take the less intrusive approach of connecting a name to the face automatically.

In fact, when it comes to identifying information, the Court has held that individuals have minimal, if any, expectation of privacy in the secrecy of their identifying information.[23]

This is because the Fourth Amendment provides no protection for what "a person knowingly exposes to the public, even in his own home or office."[24] The physical characteristics of a person's face, its shape and contours are constantly exposed to the public, so no person can have a reasonable expectation that others will not observe his facial features. Indeed, the Court has held that compelling a suspect to provide physically identifying information—such as fingerprints[25] or voice exemplars[26] —is usually reasonable because such techniques intrude upon no cognizable individual interest. A facial scan is far less intrusive than either of these. Lower courts have held similarly in other identification and location-determination cases, even including the use of beepers to pinpoint location.[27] In short, in each of these identifying-information cases, the Court has held that the relevant individual interest at stake is in personal security or privacy as seclusion, but has not acknowledged a privacy interest in keeping personal information a secret.

The measure of personal intrusion during an investigative stop using facial recognition scanning will likely be the length of the detention and the physical intrusiveness of the search techniques employed. If by using such technology the police held an individual for longer, or required that he pose in a compromising or otherwise physically arduous manner, then *those* actions could render the search more intrusive and affect its reasonableness. But since the police can use facial recognition software unobtrusively and almost instantaneously, its use should not change the constitutional calculus of a *Terry* stop or convert an otherwise reasonable search and seizure into an unreasonable one.

C. Supreme Court Doctrine Protects Privacy as Seclusion but Not Secrecy

Real property law informs whether an individual has a reasonable expectation of privacy in secluding—i.e. restricting access to others—the property searched.[28] Facial recognition technology does not interfere with a cognizable Fourth Amendment privacy interest, such as interference with real or intellectual property rights, nor does it intrude upon personal security or movement. As such, there is no source of law upon which a reasonable expectation of privacy to object to facial recognition scanning could be grounded.[29]

Concepts of possession and property are at the core of the Fourth Amendment, as its possessive pronoun makes clear: "the right of the people to be secure in *their* persons, houses, papers, and effects." And so, from the beginning,[30] the Court has turned to property law to inform Fourth Amendment interests.[31] When the Court first encountered the modern investigative technique of wiretapping, for example, which like facial recognition enables investigators to obtain evidence without any physical interference with one's property, the Court found that no search had occurred because conversations are not tangible property or "material things" that the Fourth Amendment protects.[32]

Likewise, facial recognition technology does not implicate any property interests. Now, to be sure, the Court has subsequently extended the Fourth Amendment beyond property, as Justice Brandeis had urged in dissent. In *Katz v. United States*, the Court held that the Fourth Amendment applies to tangible and intangible interests such as private conversations,[33] because it safeguards from unreasonable search and seizure what an individual "seeks to preserve as private."[34] Justice Harlan concurred and proposed the expectation-of-privacy analysis[35] that the Court eventually adopted.[36] This privacy test holds that a Fourth Amendment search occurs when an individual has a subjective expectation of privacy, that society recognizes as reasonable, which has been invaded.[37]

Consistent with this analysis, the key Fourth Amendment question concerning facial recognition technology is whether its investigative use intrudes upon a privacy interest that society recognizes as reasonable.

Even with this expanded view of individual interests, however, an individual who is scanned in public cannot reasonably claim that facial recognition technology captures something he has sought to seclude from public view. Instead, he must argue that he has a reasonable expectation of privacy in his personal identity associated with his facial features. Under current doctrine, courts would properly reject such a claim. Despite the shift in *Katz* from purely property-based privacy protections to seclusion more generally, the Court has not recognized an independent privacy interest in the secrecy of identifying information per se.

Consequently, it is the physical intrusiveness of facial recognition technology, and not the extent to which it reveals personally identifying information, that will determine its reasonableness in a Fourth Amendment inquiry.[38] And because the technology is physically unobtrusive and does not reveal information that is secluded or otherwise hidden from public view, it is not properly characterized as a Fourth Amendment search.

Most recently, in *United States v. Jones*,[39] the Court revisited its property-invasion-as- privacy rationale, holding that the government's installation of a GPS tracking device on a suspect's vehicle constitutes a search subject to the Fourth Amendment.[40] In so doing, it left open whether the Fourth Amendment protects more than just intrusion upon seclusion. Justice Scalia, writing for the majority, emphasized that the government had "physically occupied private property for the purpose of obtaining information."[41]

Invoking Lord Camden's opinion in *Entick v. Carrington*[42] and the text of the Fourth Amendment itself, Justice Scalia reaffirmed the significance of property rights to search and seizure analysis.[43] Although acknowledging that the Court had expanded beyond a strictly property-based approach in *Katz*, the opinion nevertheless emphasized that property rights remain the central source of individual interests protected by the Fourth Amendment.[44] While the police could have obtained the same information in *Jones* without a physical trespass, and such an intrusion might still be unconstitutional under *Katz*, the facts in *Jones* did not require the Court to resolve that question.[45]

What remains after *Jones* is an incomplete picture of which individual interests beyond real property interests, if any, that the Fourth Amendment protects. At the very least, *Jones* repudiates the view that *Katz* was "a shift in

Fourth Amendment jurisprudential paradigms from a property-based framework to an expectation-of-privacy framework."[46]

Real property law remains central to Fourth Amendment individual interests. And under a property analysis, facial recognition is clearly not a search.

To be sure, the *Jones* majority also emphasized that trespass upon property and the *Katz* expectation-of-privacy test coexist in Fourth Amendment jurisprudence. But even under a privacy-based analysis, the use of facial recognition technology is not analogous to the wiretap in *Katz*, and as such is not properly characterized as a search. If facial recognition technology captures facial features that an individual has not secluded,[47] then unlike *Katz*, its investigative use does not intrude upon information that an individual has kept hidden. Without trespass upon real property, or upon information that a person has sought to hide, there is no legitimate source of law upon which a reasonable expectation of privacy could be founded.

CONCLUSION

As I have explained, governmental use of facial recognition technology does not generally constitute a Fourth Amendment search, let alone an unreasonable one. Nevertheless, this technology does raise novel privacy concerns, which are certainly the proper concern of Congress.

As some of the other panelists have emphasized, it is a brave new world, in which our reasonable expectation of privacy may seem to be under assault by technology. And Congress may indeed have a role to play in striking the proper balance between privacy and security in this area. I would emphasize, though, that the legal and ethical landscape here is very complex, implicating a variety of cross-cutting interests. (The answers may be different depending on whether these tools are being used by governments or by private actors, and it may matter whether people have opted into a database or been included against their will.). These sorts of technological advances may indeed diminish our privacy. But they are also extraordinarily useful, to private individuals, to corporations, to social networks, and to law enforcement. I am not at all certain that legislation is required in this area. But in any case, I would encourage the Subcommittee to consider carefully the legal, ethical, and social implications of any legislative response. Again, I thank you for the opportunity to appear before you today, and I look forward to your questions.

End Notes

[1] Chandrakant D. Patel et al., Biometrics in IRIS Technolgy: A Survey, 2 International Journal of Scientific and Research Publications 3 (2012), *available at* (http://www.ijsrp.org/print-journal/ijsrp-jan-2012- print.pdf#page=5) (last accessed July 14, 2012).

[2] Kevin Bonsor & Ryan Johnson, *How Facial Recognition Systems Work*, HOWSTUFFWORKS.COM (http://electronics.howstuffworks.com/gadgets/high-tech-gadgets/facial-recognition.htm) (last visited Dec. 4, 2011).

[3] *See* SUBCOMM. ON BIOMETRICS, NAT'L SCI. & TECH. COUNCIL, PRIVACY & BIOMETRICS: BUILDING A CONCEPTUAL FOUNDATION 4 (2006), *available at* http://www. biometrics.gov/docs/privacy.pdf (defining biometrics as "automated methods of recognizing an individual based on measurable biological . . . and behavioral characteristics").

[4] *See id.* at 15-17 (describing existing technologies for such biometric methods).

[5] Emily Steel, *How a New Police Tool for Face Recognition Works*, WALL ST. J. DIGITS BLOG (J uly 13, 2011, 7:56 AM), http://blogs.wsj.com/digits/2011/07/13/how-a-new-police-tool-for-face-recognition-works.

[6] *See* Nita A. Farahany, *Incriminating Thoughts*, 64 STANFORD L. REV. 351, 368-70 (2012).

[7] *Id.* at 368.

[8] *Id.*

[9] U.S. CONST. amend. IV.

[10] Whether an expectation of privacy is reasonable or not has always turned on bodies of law outside of the Fourth Amendment. Privacy expectations are reflected in laws or societal norms, so a reasonable expectation of privacy "must have a source outside of the Fourth Amendment, either by reference to concepts of real or personal property law or to understandings that are recognized and permitted by society." *See* United States v. Jones, 132 S. Ct. 945, 951 (2012) ("[O]ur very definition of 'reasonable expectation of privacy' . . . we have said [is] an expectation 'that has a source outside of the Fourth Amendment, either by reference to concepts of real or personal property law or to understanding that are recognized and permitted by society.'" (quoting Minnesota v. Carter, 525 U.S. 83, 88 (1988))).

[11] *See* California v. Greenwood, 486 U.S. 35 (1988) (holding that the Fourth Amendment does not prohibit the warrantless search and seizure of garbage left for collection outside the curtilage of a home).

[12] *See* Kyllo v. United States, 533 U.S. 27 (2001) (finding the use of thermal imaging to scan a person's home to be a Fourth Amendment "search").

[13] E.g., Florida v. Riley, 488 U.S. 445 (1989) (holding that no Fourth Amendment search had occurred when a police helicopter flew overhead at low altitude a greenhouse and the pilot looked through a hole in the roof of the greenhouse and saw drugs being grown inside); California v. Ciraolo, 476 U.S. 207 (1986) (finding that aerial surveillance of a person's yard blocked from view by an outer fence because the Fourth Amendment does not require that police "shield their eyes when passing by a home on public thoroughfares," so the surveillance was not a search when it took place from a public place); Dow Chemical v. United States, 476 U.S. 227 (1986) (determining that the taking of photography from navigable airspace was not a Fourth Amendment search).

[14] *See* generally Orin S. Kerr, *An Equilibrium Theory of the Fourth Amendment*, 125 HARV. L. REV. 476, 522-29 (2011) (discussing Fourth Amendment "open fields" cases that allow surveillance of the curtilage of one's home).

[15] *See* Nita A. Farahany, *Searching Secrets*, 160 U. Penn. L. Rev. 1239, 1255 (2012), citing *Payton v. New York*, 445 U.S. 573, 585 (1980) ("[P]hysical entry of the home is the chief evil against which the Fourth Amendment was directed." (quoting United States v. U.S. Dist. Court, 407 U.S. 297, 313 (1972))); Mincey v. Arizona, 437 U.S. 385, 391 (1978) (holding that one does not forfeit her Fourth Amendment rights to her home by committing a crime); Alderman v. United States, 394 U.S. 165, 171-72 (1969) (linking property rights to the ability to raise a motion to exclude evidence based upon the Fourth Amendment).

[16] See, e.g., Jack M. Balkin, Essay, *The Constitution in the National Surveillance State*, 93 MINN. L. REV. 1, 20 (2008) ("The Fourth Amendment does not require governments to discard any information they have already lawfully collected.")

[17] *See* Nita A. Farahany, *Searching Secrets*, 17, at 1277-82 (discussing Fourth Amendment doctrine concerning identifying information); Nita A. Farahany, *Incriminating Thoughts*, supra note 6, at 368-72 (discussing the category of identifying evidence and constitutional protections of the same).

[18] 392 U.S. 1 (1968).

[19] Hiibel v. Sixth Judicial District Court of Nevada, 542 U.S. 177 (2004).

[20] *Id.* at 186-87 (citation omitted).

[21] *See id.* at 188 (noting that the "stop and identify" statute at issue in the case served the useful purpose of increasing the likelihood that a suspect would actually disclose his identity to a police officer).

[22] For example, the Court in *United States v. Dionisio*, 410 U.S. 1 (1973), noted that the Fourth Amendment provides no protection for what "a person knowingly exposes to the public, even in his own home or office . . . ." The physical characteristics of a person's voice, its tone and manner, as opposed to the content of a specific conversation, are constantly exposed to the public. . . . No person can have a reasonable expectation that others will not know the sound of his voice . . . . *Id.* at 14 (first alteration in original) (citation omitted) (quoting Katz v. United States, 389 U.S. 347, 351 (1967)).

[23] *See* Farahany, supra note 17.

[24] *United States v. Dionisio*, 410 U.S. 1, 14 (1973) (describing the reasonable expectation that others will be familiar with one's physical features).

[25] *See* Davis v. Mississippi, 394 U.S. 721, 727 (1969) ("Detention for fingerprinting may constitute a much less serious intrusion upon personal security than other types of police searches and detentions. Fingerprinting involves none of the probing into an individual's private life and thoughts that marks an interrogation or search.").

[26] *See Dionisio*, 410 U.S. at 14 ("No person can have a reasonable expectation that others will not know the sound of his voice, any more than he can reasonably expect that his face will be a mystery to the world.").

[27] *See* Recent Development, *Who Knows Where You've Been?: Privacy Concerns Regarding the Use of Cellular Phones as Personal Locators*, 18 HARV. J.L. & TECH. 307, 314-15 (2004) (describing *United States v. Forest*, 355 F.3d 942 (6th Cir. 2004), which held that pinpointing one's location or movement through one's beeper is not subject to a reasonable expectation of privacy).

[28] Richard A. Posner, *Privacy, Secrecy, and Reputation*, 28 BUFF. L. REV. 1, 3-4 (1979); *see also id.* ("An equivalent term is 'retirement' in its complex modern sense in which we speak of a person being 'retiring' and also of a person being 'retired.'").

[29] *See* Farahany, supra note 15, at 1254 (explaining that the most consistently recognized subjective and objective expectation of privacy is one that derives, at least in part, "'from the right to exclude others from the property in question.'")(internal citations omitted).

[30] U.S. Const. amend IV ("The right of the people to be secure in their persons, houses, papers, and effects, against unreasonable searches and seizures, shall not be violated.").

[31] *See* Boyd v. United States, 116 U.S. 616, 627 (1886) (stating that the "sacred and incommunicable" right of property is only set aside "for the good of the whole" (quoting Entick v. Carrington, (1765) 19 How. St. Tr. 1029 (C.P.) 1066 (Eng.)). *See also* Farahany, supra note 17, at 1244-49 (reviewing how property law has informed Fourth Amendment privacy interests).

[32] Olmstead v. United States, 277 U.S. 438, 457 (1928), *overruled by* Katz v. United States, 389 U.S. 347 (1967), *and* Berger v. New York, 388 U.S. 41 (1967).

[33] In *Katz v. United States*, 389 U.S. 347 (1967), FBI agents attached a device to the outside of a public telephone booth to listen to the defendant's conversations, and the Government argued that this eavesdropping did not implicate the Fourth Amendment because no trespass upon the defendant's property occurred. The government rejected the idea that an intrusion upon a constitutionally protected area must occur for the Fourth Amendment to apply.

[34] *Id.* at 351-52.

[35] *See id.* at 361 (Harlan, J., concurring).

[36] *See* Smith v. Maryland, 442 U.S. 735, 740-41 (1979) (The Court held that the government's use of a pen register—a device that records the phone numbers one dials—was not a Fourth Amendment search. The Court explained that "a pen register differs significantly from the listening device employed in *Katz*, for pen registers do not acquire the *contents* of communications.")

[37] *Katz*, 389 U.S. at 361. Since *Katz*, Fourth Amendment law has also addressed the concealment of information. But even in *Katz*, the Court ultimately focused on Katz's seclusion of himself in the phone booth and not on his interest in the substantive secrecy of his conversation.

[38] *See* Terry v. Ohio, 392 U.S. 1, 30-31 (1968) (holding that the Fourth Amendment permits "reasonable inquiries" to determine a suspect's identity).

[39] 132 S. Ct. 945 (2012).

[40] *Id.* at 949 (2012).

[41] *Id.*

[42] (1765) 19 How. St. Tr. 1029 (C.P.) (Eng.).

[43] *Jones*, 132 S. Ct. at 949.

[44] *Id.* at 950.

[45] *Id.* at 954. Since the government took the position that GPS tracking did not constitute a search, the Court left for another day the further question of which individual interests, beyond intrusion upon property, an individual could claim to assess the reasonableness of the search that had occurred.

[46] David A. Sullivan, A Bright Line in the Sky?: Toward a New Fourth Amendment Search Standard for Advancing Surveillance Technology, 44 ARIZ. L. REV. 967, 974 (2002).

[47] E.g. by wearing a mask, a veil, or other head covering that hides their visage.

In: Facial Recognition Technology ...
Editor: Meena N. Harnois
ISBN: 978-1-62417-572-5
© 2013 Nova Science Publishers, Inc.

Chapter 5

TESTIMONY OF ALESSANDRO ACQUISTI, ASSOCIATE PROFESSOR, CARNEGIE MELLON UNIVERSITY. HEARING ON "WHAT FACIAL RECOGNITION TECHNOLOGY MEANS FOR PRIVACY AND CIVIL LIBERTIES"*

Chairman Franken, Ranking Member Coburn, and Members of the Subcommittee: I am honored to appear before you today. I am a tenured associate professor at the Heinz College, Carnegie Mellon University (CMU), a member of CMU CyLab, and the co-director of CMU's Center for Behavioral Decision Research (CBDR).[1] I am an economist by training (I hold Master degrees in economics from Trinity College Dublin and from the London School of Economics), and I am interested in how economics can help us understand the impact of information technology (I hold a PhD in Information Management and Systems from the University of California at Berkeley). For about 10 years, I have been studying the economics and behavioral economics of privacy. My research in this area has combined economics, experimental behavioral decision research, and information technology to investigate the trade-offs associated with the protection and disclosure of personal information, the technologies that enhance the former or

* This is an edited, reformatted and augmented version of testimony given on July 18, 2012 before the Senate Judiciary Committee, Subcommittee on Privacy, Technology and the Law.

the latter,[2] and how individuals value, and make decisions about, those trade-offs.[3]

My remarks in this testimony will concern research that I and others have carried out in the field of privacy and face recognition.[4] I became interested in face recognition indirectly, as a result of my studies of privacy and online social networks, which started in 2005.[5] In the summer of 2011, together with my colleagues Dr. Ralph Gross (a face recognition expert at Carnegie Mellon University) and Dr. Fred Stutzman (an online social networks expert also at Carnegie Mellon University), I presented the results of a series of experiments about the privacy implications of the convergence of more accurate face recognition technology, increasing public availability of personal data (including digital photos), and statistical re-identification techniques.[6]

In my testimony, I will highlight four conclusions from my and other scholars' research in this area:

First, face recognition is "now." The technology has evolved over several decades. While early algorithms vastly underperformed human ability to detect and recognize faces, modern ones have progressed to a point that they are now being deployed in end-user applications.

Second, the convergence of face recognition, online social networks, and data mining has made it possible to use publicly available data and inexpensive off-the-shelf technologies to produce sensitive inferences merely starting from an anonymous face. I will highlight the results of three experiments we conducted in this area, including one in which we predicted portions of the Social Security numbers of students at a North-American college starting from photos of their faces.

Third, face recognition, like other information technologies, can be source of both benefits and costs to society and its individual members. However, the combination of face recognition, social networks data, and data mining, can significantly undermine our current notions and expectations of privacy and anonymity.

Fourth, depending on which goals Congress intends to achieve in the area of face recognition, different policies and interventions may be considered. If the privacy and civil liberties implications of face recognition are the concern, it is unlikely that industry solutions alone (such as self-regulatory efforts) will address those concerns.

1. The State of Face Recognition Research

Research in computer face recognition has been conducted for over forty years.[7] For most of this time, computers underperformed humans in tasks involving the detection and recognition of individuals through their faces. But the progress of algorithms has been steady: since 1993, the error rate of face recognition systems has decreased by a factor of 272.[8] Today, under certain conditions, machine face recognition performance can be comparable or even better than humans at recognizing faces.[9]

As face recognizers' accuracy kept increasing, face recognition started being deployed in more products and services. About 10 years ago, face recognizers remained the domain of government agencies or large corporations, and were mostly used in security and police activities.[10] In the past few years, face detection, face recognition, and other related algorithms (such as the estimation of individual traits from facial images) have started appearing in end-user products, and in particular Web 2.0 services. Following the acquisition of Neven Vision in 2006 and of Like.com in 2010, Google has offered Picasa users face recognition tools to organize photos according to the individuals in them.[11]

Apple's iPhoto has employed face recognition to identify faces in a person's album since 2009.[12] Using Face.com's licensed technology, Facebook has used face recognition to suggest "tags" of individuals found in members' photos.[13] Klik, also developed by Face.com, used face recognition to allow real time tagging of Facebook friends through the same mobile camera used to take their pictures.[14]

NEC has designed billboards that automatically locate faces of people passing by and estimate their gender and age, in order to target advertising accordingly.[15] SceneTap uses face detection to estimate the size of and gender ratio in a crowd of patrons at a venue – and allows individuals who downloaded its app to find that information online.[16]

Under typical, real life conditions (for instance, when facial shots are not captured in well-lit conditions or through frontal, "mugshot" poses) computer face recognition still underperforms humans. As we further discuss in the next section, however, face recognition's limitations are more transient than systemic: given the growing commercial interest in face recognition and its application, the gap between humans and machines in recognizing faces is likely to keep diminishing.

2. How to Predict Someone's SSN Starting from Their Face

As a privacy researcher, a few years ago I became interested in the privacy implications of the convergence of two trends: the improving ability of computer programs to recognize faces in digital images, and the increasing public availability of identified facial photos online - especially through online social networks. In a study with Ralph Gross and Fred Stutzman presented in the summer of 2011,[17] we investigated whether the combination of publicly available Web 2.0 data and off-the-shelf face recognition software may allow large-scale, automated, end-user individual re-identification. We identified individuals online across different services (Experiment 1); offline – that is, in the physical world (Experiment 2); and then inferred additional, sensitive information about them (their interests and their Social Security numbers), combining face recognition and data mining, thus blending together online and offline data (Experiment 3); finally, we developed a mobile phone application to demonstrate the ability to recognize and then predict someone's sensitive personal data directly from their face, in real time, on a mobile device.

In the first experiment (Experiment 1) we investigated online-to-online re-identification. We took unidentified profile photos from a popular dating site (where people use pseudonyms to protect privacy), compared them - using face recognition - to identified photos from a popular online social network (Facebook; namely, we used the component of a Facebook profile that can be publicly accessed via a search engine; in other words, we did not even need to log on to the network itself). Through this process, we were able to re-identify about 10% of the pseudonymous members of the dating site.

In the second experiment (Experiment 2) we investigated offline-to-online re-identification. Its methodology was conceptually similar to that of Experiment 1, but in this case we attempted to re-identify students on the campus of a North American college. We took photos of them with a webcam and then compared those shots to images from Facebook profiles. Using this approach, we re-identified about one third of the subjects in the experiment.

The first two experiments illustrate how third parties can use publicly available data not just for contextual identification (linking images of the same person in an album of photos) but also for universal, unique identification.[18] In the final experiment (Experiment 3), we predicted the interests and the first five digits of Social Security numbers (SSNs) of some of the individuals who had participated in the second experiment. We did so by combining face recognition with the algorithms we developed in 2009 to predict SSNs from public data.[19]

In the context of Experiment 3, SSNs were merely one example of the many types of information it is possible to infer about a person, starting merely from an anonymous face, through a chain of inferences in a process of *data accretion*.[20] The process illustrates the privacy implications of face recognition technology, and can be summarized in the following manner: First, face recognition links an unidentified subject (for instance, a face among many in the street) to a record in an identified database (such as an identified photo of the subject on Facebook, LinkedIn, Amazon, or in a state's DMV database). Once the link has been established, any online information associated with that record in the identified database (such as names and interests found in the subject's Facebook profile; or demographic data found on Spokeo.com - a social network and data aggregator) can in turn be probabilistically linked to the unidentified subject. Lastly, through data mining and statistical re-identification techniques, such online information can be used for additional, and much more sensitive inferences (such as sexual orientation,[21] or Social Security numbers[22]), which, in turn, can be linked back to the originally unidentified face. Sensitive data is therefore linked to an anonymous face through what we may refer to as a "transitive property" of (personal) information - a process that merely requires publicly available data. Sensitive information thus becomes "personally predictable information."

As a further example of what is already possible to accomplish using existing technologies and publicly available data, we developed a mobile phone application which, once a photo is taken of a person's face, uploads it to a server; there, a program compares it to a database of images downloaded from the Internet – and tries to recognize the person; thereafter, using the same process adopted in Experiment 3, the program attempts to predict the person's sensitive personal information, and finally displays that very same information on the device's screen, overlaid on the face of the subject.

Essentially, this application (which we demoed at a security conference in August 2011, with no intention of making it publicly available) demonstrates that it is possible to conduct on a mobile device and in real time the process of Experiment 3.

We use the term *augmented reality* to describe that application: it refers to the merging of online and offline data that new technologies make possible. If an individual's face on the street can be identified using a face recognizer and identified images from social network sites such as Facebook or LinkedIn, then it becomes possible not just to identify that individual, but also to infer additional, and more sensitive, information about her.

2.1. Current Limitations and Why They Are Transient

Our experiments, while successful, were constrained geographically (we focused on the population of a North-American city, and the students at a North-American college campus) and in scope (we used databases with up to a few hundred thousands of images). The results we obtained are not *yet* scalable to the entire American population for a number of reasons: First, computational costs: we estimate that comparing the shot of a person's face to a database with mugshots of 280 million US residents aged 14 years or older, using the same hardware as in our experiments, would take over four hours (rather than the few seconds that process took in our experiments). Second, "false positives:" when comparing millions of human faces, several individuals' faces will be similar to each other, and computers do not yet excel in separating a face of a person from a face of someone who looks very much like that person. Third, light conditions, facial hair, or non-frontal poses impair the accuracy of machine face recognizers. Fourth, photographic images (and in particular frontal mugshots) may not be available for the entire population.

Those hurdles, however, are being progressively overcome. They are transient, not systemic. First, as computers' processing power keeps increasing, and cloud computing costs decrease, it will become more efficient to run end-user applications on mobile devices similar to the one we developed, for mass-scale, automated, peer-to-peer face recognition. Second, false positives will likely keep reducing, as machine face recognition error rates are decreasing by one half about every two years.[23]

Third, researchers are actively working on improving computers' ability to recognize faces under varied conditions of light, facial hair, and poses. Fourth, entities such as online social networks are amassing some of the largest known databases of identified photos, from which increasingly accurate biometric models or "faceprints" of increasing portions of the US population can be, and are being, built.

It is likely, therefore, that within a few years, real-time, automated, mass-scale facial recognition will be technologically feasible and economically efficient. It will be feasible for individual end-users, in a peer-to-peer fashion; it will be feasible for firms (both for companies, such as online social networks, that will actually own the data; and for third parties that will rent identity recognition services from or through the former); and it will be feasible for governments that will access or trade biometric data with the private sector. In this world, "facial searches" in the street may become as common as text-based queries on search engines are today.

However, the fact that real time recognition technology may be used, in principle, by individuals, firms, and governments alike, does not guarantee that, in practice, all parties will gain equal access to it. The parties in actual possession of the largest databases of images will be in the better position to make use of face recognition technology and control the access that others will have to it. Similarly, the parties with more resources will be those more likely to be able to exploit it, or avoid being exploited by it.

3. Trade-offs and Privacy Concerns

In economic terms, privacy shares the characteristics of both an *intermediate good* (a good that is valued in an instrumental way, because of its consequences: for instance, loss of personal data can cause identity theft and ensuing economic damage), and a *final good* (a good that is valued for its own sake: for instance, ubiquitous surveillance creates discomfort).[24]

I will first discuss the implications of face recognition for privacy as an intermediate good. As it is often the case with advances in information technology, the ramifications of cheap, powerful facial recognition technology in the hands of individuals, firms, and government agencies, are complex. They include both scenarios where public or individual welfare are increased, and scenarios where significant tangible and intangible costs arise. Consider a peer-to-peer scenario: Your phone (or in some years your glasses, and in a few more your contact lenses) will tell you the name of that person at the party whose name you always forget; or, it will tell the stalker in the bar the address where you live. Consider a third party firm scenario: The hotel will recognize and greet you as you enter the lobby with your luggage (because you friended them on a social network, or because - with or without your consent - the hotel enrolled in some identity recognition service sold by a social network); or, the salesperson will infer your credit score the moment you enter the dealership, and use a psychological profile (also calculated in real time from your online posts) to nudge you to accept a steep price for the car you wanted. Consider a government agency scenario: An investigative agency will be able to find missing or exploited children in databases of online photos; or, an administration will be able to identify from remote, high-definition cameras, all of the thousands of participants in a peaceful protest. More scenarios are unforeseeable today but will be commonplace tomorrow.

In short, face recognition could make our lives easier, or more comfortable, or more secure; conversely, it could limit our freedom, endanger our security, ease the extraction of consumer surplus, and chill free speech by creating a state of constant and ubiquitous surveillance.

Next, I will discuss the implications of face recognition for privacy as a final good: in this regard, there are reasons to believe that the process through which face recognition erodes our notion of privacy has not just started, but is well on its way. Note that the results of our experiments were limited by design: our self-imposed constraint consisted in only using publicly available data and technologies that other third parties and end-users could also get access to: off-the-shelf face recognition technology,[25] limited computational power accessed through cloud computing services, and limited amounts of facial images made publicly available by end users on online social networks. In reality, today, both governmental and private sector entities have access to more powerful computational tools and much larger (and more accurate) repositories of digital photos than we had.

Consider, for instance, Facebook – currently the largest social networking site. Many of its users (estimated at over 900 million worldwide,[26] with more than 90 billion of photos allegedly collectively uploaded[27]) choose photos of themselves as their "primary profile" image. These photos are often identifiable: Facebook has aggressively pursued a "real identity" policy, under which members are expected to join the network under their real names, under penalty of account cancelation.[28] Using tagging features and login security questions, the social network has successfully nudged users to associate their and their friends' names to uploaded photos. These photos are also publicly available: Primary profile photos *must* be shared with strangers under Facebook's own Privacy Policy.[29] Many members also allow those photos to be searchable from outside the network via search engines.

Online social networks such as Facebook are accumulating the largest known databases of facial images. Often, those images are tagged or attached to fully identified profiles, thus providing a linkage between a person's facial biometrics and their real names. Furthermore, many social network users post and tag *multiple* photos of themselves - and their friends, allowing biometric models of their faces, and those of other people as well, to become more accurate. Before Face.com was acquired by Facebook, for instance,[30] its mobile face recognition application Klik effectively used its own users as means of improving the recognition accuracy of its algorithms (since users were asked to select the correct name among a list of possible matches found by the application for a given face). This process would increase the future recognizability even of subjects who did not explicitly consent to more accurate biometric models of their faces being composed, or who had no knowledge of that happening.[31] Furthermore, such vast and centralize biometrics database can be at risk of third-party hacking.[32]

Online social networks are becoming today's de facto "Real IDs" – produced not by legislative mandate but technological capability. We must realize that the notions and expectations of privacy and anonymity, as we have known them for the history of human kind – the idea that, among strangers, you are a stranger, are facing a new challenge that we do not yet fully comprehend.

4. IMPLICATIONS

The evolution of face recognition technology is inescapable, but its applications will not all be equally desirable. Unfortunately, there is no obvious silver bullet that will allow society to continue to enjoy the benefit of face recognition divorced from its more concerning usages. There are, however, policy and technology mechanisms that are worth of investigation, and which carry trade-offs we are only beginning to appreciate: privacy enhancing technologies applied to face capture, detection, or recognition;[33] moratoriums on specific applications of face recognition;[34] explicit consent requirements to having one face's tracked, or matched, in public; do-not-recognize-me mechanisms; or privacy legislation that would place obligations on those who use facial recognition to collect the personal identity and thereby the personal information of others; and so forth. Unfortunately, the mere reliance on industry self-regulation is unlikely to find balance between uses and abuses of face recognition, due to the particular economic value of identified facial information, and recent history in the markets for personal data.

Identified facial information is especially valuable because facial biometric models are peculiarly sensitive and powerful instruments of identification and tracking. First, a person's face is a permanent identifier: it changes over time in patterns that computers are learning to predict, but it cannot be permanently altered to avoid detection without great cost.[35] Second, a person's face is a veritable conduit between her different persona: it can link a person's online world (for instance, her social network presence) to her offline world (the person walking in the street). Third, it can be captured remotely and surreptitiously, and therefore without individual consent or knowledge. Fourth, the technologies necessary to track and identify faces, as discussed, are becoming ubiquitous. As a result, control over vast repositories of identified facial information can give a firm unique power to serve and influence an array of other services and applications; competition for that control, therefore, will be fierce, at the likely cost of the privacy of end users.[36]

An analysis of recent history in the market for personal data also suggests that firms may engage in more invasive applications of face recognition over time. Currently, end-user applications of face recognition have limited capabilities. Yet those limitations are less driven by technological constraints than by firms' concerns over consumers' reaction to too-aggressive deployment of face recognition. Evidence of this can be gleaned from firms' assurances that these services do not (yet) allow indiscriminate face recognition of everybody: for instance, Face.com developed the face recognition tagging services used by Facebook users --- but "if you choose to hide your Facebook tags, [Face.com] services will get blocked out when attempting to recognize you in photos."[37] However, if recent history of privacy in social networks is a guide, the current, almost coy applications of face recognition may be "bridgeheads" designed by firms to habituate end-users into progressively more powerful and intrusive services. Consider the frequency in which, in the past few years, a popular social network such as Facebook has engaged in practices that either a) unilaterally modified settings or defaults associated with users' privacy, so as to force increased sharing or disclosure;[38] and b) reflected a "two steps forward, one step backward" strategy, in which new services were enacted or proposed, then taken back or scaled down due to users' reaction to their invasiveness, and then enacted again, after some time had passed.[39] The fact that, as consumers, we do get eventually habituated to those new services does not necessarily prove that they come without risks: Our attention is captured by what we can see as their immediate benefits; what we pay less attention to are their privacy costs, as they are often delayed.[40]

In the absence of policy intervention, therefore, the patterns we are observing (increasing gathering and usage of individuals' facial biometrics data) are unlikely to abate. The risk exists that some firms may attempt to strategically use default settings, unilateral changes to interfaces and systems, and user habituation to nudge individuals into accepting more capturing and usage of facial data – creating a perception of *fait accompli* which, in turn, will influence individuals' expectations of privacy and anonymity.

Information is power. In the 21st century, the wealth of granular data accumulated about each individual, and the staggering progress of behavioral sciences in understanding how that knowledge can be used to nudge and influence individual behavior, make it so that control over personal information will imply power over the person. It does not matter whether this control will be exercised by a government or by a corporation: as control is tilting from data subjects to data holders, the very balance of power between different entities is at stake. Senators, I do not envy your position. We had the easy task – showing the problem. You have the much harder task of helping steer us towards its solution.

Thank you for inviting me to testify today. I look forward to answering your questions.

End Notes

[1] See http://www.heinz.cmu.edu/~acquisti/

[2] A. Acquisti, 2010. "The Economics Of Personal Data and The Economics Of Privacy." Commissioned By The OECD, For The OECD Roundtable On The Economics Of Privacy and Personal Data, Paris, December 2010. http://www.heinz.cmu.edu/~acquisti/papers/acquisti-privacy-OECD-22-11-10.pdf.

[3] A. Acquisti, 2004. "Privacy In Electronic Commerce and The Economics Of Immediate Gratification." ACM Electronic Commerce Conference, 21-29.

[4] In this testimony, I use sometimes "face recognition" as a short-hand for machine-based, or computer, face recognition.

[5] R. Gross and A. Acquisti, 2005. "Information Revelation and Privacy in Online Social Networks." Proceedings of the 2005 Workshop on Privacy in the Electronic Society (WPES), ACM, 71-80, 2005; A. Acquisti and R. Gross, 2006. "Imagined Communities: Awareness, Information Sharing, and Privacy on the Facebook." Proceedings of Privacy Enhancing Technologies Workshop (PET), Lecture Notes in Computer Science 4258, Springer, 36-58.

[6] A. Acquisti, R. Gross, and F. Stutzman, 2011. "Faces of Facebook: Privacy In The Age Of Augmented Reality." Proceedings of Blackhat USA. http://www.heinz.cmu.edu/~acquisti/face-recognition-study-FAQ/.

[7] Some of the earliest works in this area include, e.g., W.W. Bledsoe, 1964. "The model method in facial recognition." Tech. rep. PRI:15, Panoramic Research Inc., Palo Alto, CA; T. Kanada, 1973. "Computer recognition of human faces." Birkhauser, Basel, Switzerland, and Stuttgart, Germany; M.D. Kelly, 1970. "Visual identification of people by computer." Tech. rep. AI-130, Stanford AI Project, Stanford, CA.

[8] P.J. Phillips, 2011. "Improving Face Recognition Technology" Computer, 44(3), 84-86.

[9] For instance, computers outperform humans in recognizing unfamiliar faces: P.J. Phillips et al, 2007. "FRVT 2006 and ICE 2006 large-scale results." National Institute of Standards and Technology, #7408.

[10] For instance, significant debate accompanied the usage of face recognition during Super Bowl 2001; see http://www.wired.com/politics/law/news/2001/02/41571.

[11] See http://picasa.google.com/support/bin/answer.py?answer=156272. More recently (in the summer of 2011) Google also acquired Pittsburgh-based face recognition company PittPatt.

[12] See http://support.apple.com/kb/ht344.

[13] See http://www.facebook.com/blog.php?post=403838582130. More recently (in the summer of 2012), Facebook has acquired Face.com.

[14] See http://www.face.com. The application is no longer available to end-users after Face.com was acquired by Facebook.

[15] See http://www.guardian.co.uk/media/pda/2010/sep/27/advertising-billboards-facial-recognition-japan.

[16] See http://www.scenetap.com.

[17] A. Acquisti, R. Gross, and F. Stutzman, 2011. "Faces of Facebook: Privacy In The Age Of Augmented Reality." Proceedings of Blackhat USA. http://www.heinz.cmu.edu/~acquisti/face-recognition-study-FAQ/.

[18] See S. Garfinkel and B. Rosenberg, 2009. "Face Recognition: Clever or Just Plain Creepy?" MIT Technology Review, February 27.

[19] A. Acquisti and R. Gross, 2009. "Predicting Social Security Numbers From Public Data." Proceedings Of The National Academy Of Science, 106(27), 10975-10980.

[20] P. Ohm, 2010. "Broken Promises of Privacy: Responding to the Surprising Failure of Anonymization." UCLA Law Review, 57, 1701.

[21] C. Jernigan and B.F.T. Mistree, 2009. "Gaydar: Facebook friendships expose sexual orientation." First Monday,14(10).

[22] A. Acquisti and R. Gross, 2009. "Predicting Social Security Numbers From Public Data." Proceedings Of The National Academy Of Science, 106(27), 10975-10980.

[23] P.J. Phillips, 2011. "Improving Face Recognition Technology" Computer, 44(3), 84-86.

[24] J. Farrell, 2011. "Can Privacy Be Just Another Good?" Remarks presented at University of Colorado at Boulder's Conference on the Economics of Privacy.

[25] We used PittPatt, a face recognition application developed by former CMU researchers. Just a few days before our results were publicly presented, PittPatt was acquired by Google, thus the technology is no longer publicly available.

[26] See http://www.facebook.com/press/info.php?statistics.

[27] See http://www.quora.com/How-many-photos-are-uploaded-to-Facebook-each-day.

[28] See http://www.guardian.co.uk/world/2011/mar/09/chinese-blogger-mark-zuckerberg-dog.

[29] Consider: "Facebook is designed to make it easy for you to find and connect with others. For this reason, your name and profile picture do not have privacy settings," from http://www.facebook.com/policy.php, accessed July 22, 2011.

[30] See http://news.cnet.com/8301-1023_3-57455287-93/facebook-acquires-face.com-for-undisclosed-sum/.

[31] At the time of writing, one's images on Facebook by default can be tagged by people in their network, unless the individual explicitly opts-out. The power of default settings in nudging people's privacy choices has been explored, among others, by A. Acquisti, L. John, and G. Loewenstein, 2010. "What Is Privacy Worth?" In Workshop On The Economics of Information Security. (Leading Paper, 2010 Future Of Privacy Forum's Best "Privacy Papers For Policy Makers" Competition.) http://www.heinz.cmu.edu/~acquisti/papers/acquisti-ISR-worth.pdf. Furthermore, frequent users' errors in choosing privacy settings have been found by A. Acquisti and R. Gross, 2006. "Imagined Communities: Awareness,

Information Sharing, and Privacy on the Facebook." Proceedings of Privacy Enhancing Technologies Workshop (PET), Lecture Notes in Computer Science 4258, Springer, 36-58.

[32] For instance, security researcher Ashkan Soltani exploited a vulnerability in Klik to gain access to non-public photos and other potentially private data of Facebook users. See F. Rashid, 2012. "Face.com Fixes Facebook Hijacking Flaw in KLIK Mobile App." SecurityWatch, PCMag.com, June 20, 2012.

[33] Research in this area has been carried out by, among others, Ralph Gross and Latanya Sweeney.

[34] See http://epic.org/2012/02/epic-calls-for-moratorium-on-f.html.

[35] Consider, for instance, plastic surgery. We are not referring to temporary solutions (such as masks and cosmetic make-up), whose costs increases with the amount of time the person has to carry them in the course of their lives.

[36] While studies have shown that a significant portion of consumers are willing to pay price premia to purchase from more privacy protective merchants (J. Tsai, S. Egelman, L. Cranor, and A. Acquisti, 2011. "The Effect Of Online Privacy Information On Purchasing Behavior: An Experimental Study." Information Systems Research, 22, 254-268), it is not obvious that competition alone stimulates a market for privacy products. See, for more details. A. Acquisti, 2010. "The Economics Of Personal Data and The Economics Of Privacy." Commissioned By The OECD, For The OECD Roundtable On The Economics Of Privacy and Personal Data, Paris, December 2010.

[37] Extracted from Face.com FAQ in July 2011.

[38] Consider examples such as Facebook News Feed in 2006; Tagging in 2009; the changes in privacy settings in late 2009/early 2010; the lifting of cache time limits in 2010 (previously, user data accessed through APIs could only be cached by developers for 24 hours); the introduction of Facebook Places in 2010 (which allows others to tag a user in a certain location); the compulsory switch to "Timeline" in late 2011/early 2012; and the more recent switching of users to Facebook emails. These examples are discussed in work in progress by Alessandro Acquisti, Fred Stutzman, and Ralph Gross, and have been analyzed by research assistants at Carnegie Mellon University Seth Monteith and Nikolas Smart.

[39] Examples (also discussed in the work in progress mentioned in the previous footnote), include Beacon/Connect (2007-2008); and ToS change relative to the closing of a Facebook account in 2009.

[40] A. Acquisti, 2004. "Privacy In Electronic Commerce and The Economics Of Immediate Gratification." ACM Electronic Commerce Conference, 21-29.

In: Facial Recognition Technology ... ISBN: 978-1-62417-572-5
Editor: Meena N. Harnois © 2013 Nova Science Publishers, Inc.

Chapter 6

STATEMENT OF BRIAN MARTIN, DIRECTOR OF BIOMETRIC RESEARCH, MORPHOTRUST USA. HEARING ON "WHAT FACIAL RECOGNITION TECHNOLOGY MEANS FOR PRIVACY AND CIVIL LIBERTIES"*

Good afternoon Chairman Franken, Senator Coburn, and other distinguished members of the Subcommittee. Thank you for asking MorphoTrust USA to discuss the capabilities of facial recognition technology today.

My name is Dr. Brian Martin and as the Director of Biometric Research for MorphoTrust USA, my primary responsibility is the research and development of our company's biometric search engine. After earning a Ph.D. in Physics from the University of Pittsburgh, my career in biometrics began at a startup company called Visionics, which pioneered software-based face recognition technologies.

Over the last 15 years, my teams have played an integral part in the research and development of world class algorithms and search engines for face, iris, and fingerprint matching. These technologies are used by the U.S. Departments of Defense and State, the FBI, over 30 face recognition-enabled

* This is an edited, reformatted and augmented version of testimony given on July 18, 2012 before the Senate Judiciary Committee, Subcommittee on Privacy, Technology and the Law.

drivers' license systems, and by several large international biometric systems including the world's largest in India. With a decade and a half of real world experience in developing biometric systems, I will address the following in my testimony:

- An explanation of how facial recognition algorithms work;
- A statement on the accuracy and limitations of current state-of-the-art face recognition technologies;
- An overview of the different categories of face recognition applications; and
- Some comments on face recognition technology/design as it relates to privacy.

MorphoTrust USA's History and Role in Identity Solutions

MorphoTrust USA was formed when L-1 Identity Solutions was acquired in July 2011 by Safran, a global technology powerhouse in aerospace, defense, and security. Headquartered in Billerica, Massachusetts, MorphoTrust has over 1,100 employees across the country, including a biometrics facility in Bloomington, Minnesota.

MorphoTrust USA is the leading domestic provider of U.S. driver licenses, passports and passport cards. We provide solutions for border management, public safety, law enforcement, retail, travel and applicant vetting.

We develop the technology for and deliver some of the largest, most complex biometric systems in the world, which are used for searching large databases to prevent identity fraud, provide criminal investigative leads, and fight terrorism.

Our accomplishments range from introducing the first face recognition powered de-duplication of driver's license databases to providing the first commercial face detection technology to digital camera manufacturers. Under previous names, Visionics, Identix, Viisage, and L1 Identity Solutions, we have been at the forefront in the adoption of face recognition systems used by states and the federal government for over a decade.

How Face Recognition Works

Automated face recognition algorithms were first studied in the late 1980's and became popular in the mid 1990's. Over the last 20 years, the technology has matured to the point where it can be used as a tool to help prevent identity fraud, to provide leads in criminal investigations, and fight terrorism. The technology is based on pattern recognition techniques used in the field of computer vision.

Though there are several different approaches to face recognition, each with its own merits, most modern commercial grade algorithms follow these general steps:

1. *Detection:* First, patterns in an image are extracted and compared to, or tested against, a model of a face. When these patterns are determined to closely resemble the face model, the assumption is that a face is present in the image. This is called face detection, and in itself, can be a challenging research problem due to the large variability in what a face could look like in an image. Changes in the pose of the face, the expression on the face and the lighting (shadows) on the face make what is seemingly trivial for the human brain to accomplish an active area of research for computer vision scientists. Furthermore, algorithms have to discriminate between items that look similar to faces, but are not human faces (think of how one sees the man in the moon).
2. *Registration:* The next step following face detection is called 'feature registration'. The algorithms focus on the area of the image where a face was detected and attempt to determine the locations of a common set of facial features that will be used as key points when extracting the binary template or faceprint. The most commonly used registration points are the center of the eyes, but algorithms can use others, such as the tip of the nose, the corners of the mouth, etc. Once the algorithm determines the location of these points of interest, the features of the face are said to be registered or localized. If the algorithm cannot find suitable features in this stage, feature registration will fail, causing the face to 'fail to enroll' in the system. Note that a human can aid in these first two steps, and in fact, face detection and feature registration were performed manually in early face recognition algorithms.

3. *Feature Extraction:* Once the features are registered, various forms of image processing can be performed to normalize the image, reduce noise in the image, reduce lighting and expression variations, and even normalize the pose of the face. This image processing helps to remove variations in the face that the matching algorithms cannot easily deal with. For instance, algorithms may not be able to match the same face in two different images if they are simply at a different scale (e.g., one is more zoomed in than the other). This stage would normalize the face in the image to ensure that it is the same size as the other faces it may be matched against. After this image processing is complete, features are extracted from the face into a binary representation appropriate for classification and/or matching. This is often referred to as a facial template or a faceprint. The feature extraction step is usually quite complex and can vary drastically from algorithm to algorithm. That is, the faceprint from one approach is rarely if ever compatible with a different approach or implementation.
4. *Classification:* An optional step is face classification. With the faceprint in hand, algorithms can be trained to classify the face into any number of categories that can be used to aid face matching or can just be informational. Some examples would be to use a classifying algorithm to estimate the gender or age of the face or even estimate if the extracted features of the face are of sufficient quality to support an accurate match of the face.
5. *Matching:* Finally, after the features of a face have been generated for two presentations of a face, an algorithm can be applied to match the two faceprints against each other to produce a single score value that represents the amount of similarity between the two faces. Depending on the features used and the efficiency of the representation of the features, the complexity of the match can be extremely high and CPU intensive taking on the order of a second per match or the complexity can be very low allowing 10's of millions of matches per second on a modern server computer. An example of a simple matching algorithm would be one that simply counts how many times the 1's and 0's are different between two binary faceprints. When the count of differing bits is low, the two faces are given a higher similarity (match) score compared to a case when the counts of differing bits are high.

Though the general recipe for face recognition is similar for many approaches, the details can vary dramatically. For example, the facial features used for matching could be texture-based, such as the pattern of a hairline or eyebrow, or the features could be shape-based, where the curvature of facial features is used for matching.

3D features can be estimated from the information in an image or can be directly measured from the image capture system. Features used for matching can be global, such as the shape of the head; they can be local, such as the shape of the eye, or they can be nearly microscopic such as the pores and wrinkles in the skin. In all cases, the approaches are vastly more complicated than the commonly perceived notion that face recognition systems simply look at geometrical distance measures between local features of the face such as the eyes, nose, and mouth.

Despite the complexity, the technology is currently at a state where these face recognition algorithms can be deployed in anything from cell phones to large multi- server search engines capable of searching over 100,000,000 faces in just a few seconds with operational accuracy.

Accuracy of Facial Recognition Technology

For almost two decades, the U.S. Government has benchmarked the accuracy of automated face recognition systems. The National Institute of Standards and Technology (NIST) is currently viewed as the worldwide leader in independent benchmarking of state-of-the-art biometric technology. In NIST's 2010 face recognition report (NIST Interagency Report 7709 - Report on the Evaluation of 2D Still-Image Face Recognition Algorithms), it was shown that the best face recognition algorithms have improved by two orders of magnitude (over 100 times better) over the last decade. That is, the best algorithms can correctly determine if two faces belong to the same person 99.7% of the time while at the same time only making a mistake by falsely matching a face to the wrong person 0.1% of the time. In 1997, the algorithms could determine only if faces belonged to the same person about 50% of the time at this 0.1% false matching rate. Current state-of-the-art algorithms can in fact match faces as accurately as humans who are not trained to be experts in face matching.

This high accuracy is realized when the faces are captured in a controlled or staged environment with cooperative subjects. When the face is not looking directly at the camera, when there are strong shadows on the face, and when the image resolution is low (as one would expect from convenience store surveillance video) the accuracy of facial recognition can approach that from the late 1990's, making these scenarios an active area of current face recognition research. It is likely that recognition of faces captured in uncontrolled environments will dramatically improve over the next couple of decades, and as camera technology improves, it may allow face matching at accuracies close to what we see now in ideal conditions. Nevertheless, even when accuracy is relatively low, face recognition still proves to be a valuable tool for investigative searches. For instance, if face recognition can provide an investigative lead to a crime only 50% of the time, it is still helping to solve crimes whereas in the past it would have been unlikely to find any leads on suspects.

FACE RECOGNITION APPLICATIONS

In my earlier remarks, I addressed the concept of accuracy, but to be fair, there are different measures of accuracy for different types of applications. There are generally two main types of applications: verification and identification. The first, verification, is where the face recognition system is used to verify that you are who you say you are. This applies to the scenario of using a biometric to open a door or using your face to unlock your phone. The accuracy numbers mentioned above reflect precisely this type of application. Arguably more demanding (and more useful) is the application of identification where face recognition is used to determine an unknown identity from a gallery of known identities. This would include applications where a photo from a crime scene could be compared to a database of known offenders to generate investigative leads or where faces in the database are compared to the database itself to detect people committing identity fraud by enrolling twice under different identities.

With identification, the requirements on an algorithm are very demanding since in order to perform a single identification, the recognition system must, in an oversimplified description, perform multiple verifications – one for each member in the gallery. Identification from a database of one million faces is, in effect, the same as performing one million verifications, and consequently the algorithms should be one million times more accurate to ensure similar false positive match rates compared to the verification application. Though most modern identification systems cannot be simplified to performing several verification attempts, the point is that as the database of identities grows to the size of millions of records, the ability to perform accuracy matching becomes that much harder, and requires that much more matching power. With today's computers and state-of-the-art face recognition technology, tens of millions of records can be accurately matched per second on a single computer enabling very large scale identification applications while only taking up a relatively small hardware footprint.

Unlike verification where the system runs with very little human intervention, most large face identification systems require an expert human operator to be 'in the loop' since the chance of getting a false match in identification is directly related to the size of the gallery of faces. These expert operators either help narrow down and direct the facial search to a smaller target set of individuals or they are used to validate the potential match candidates from a facial search. This would be the case where a criminal investigator could narrow the search to suspects who live in a specific neighborhood where a crime was committed and because the pool of candidates is smaller, there is a better chance of finding a correct face matching result. Similarly, the investigator would review the recommended list of candidates from the face recognition technology after the search is complete to validate or correct the face algorithms decisions. This is just what one is doing when they correct the automatic face labels generated by photo organization software that uses face recognition.

An offshoot application of facial recognition is simply the ability to classify or characterize faces. In this application, there is no actual matching of faceprints to known or unknown identities. Instead, the features of the face are analyzed to estimate any number of aspects of the captured face. These could include gender, age, expression, at what direction the subject is looking, etc… with the most obvious use being to collect demographic information about the subjects in front of a camera. Since this can be performed in real-time with a video camera, advertisers can use this information to tailor the advertisements based on the faces looking at them by dynamically changing the content of digital billboards.

FACE RECOGNITION SYSTEM DESIGN AND PRIVACY

MorphoTrust USA face recognition search engines are designed to store and search faceprints anonymously. That is, the faceprint is identified only by a system-specific number, which the customer of the system must link to the person's identity. This makes a faceprint match, by itself, relatively useless unless the attached identity information is available. Customers that implement face recognition systems take the responsibility for the security of their identity data, since it is this customer- owned identity database that connects various metadata to the individual, such as account numbers, home address, etc. These connections to other metadata are what can be exploited in unexpected ways.

In terms of face recognition technology, the faceprints themselves contain no more information than what was in the original images from which they were derived. If a faceprint database were compromised, the faceprints could not be reverse- engineered to recreate the original face image since the faceprints are stored in a proprietary format. The faceprints are also vendor specific and are of little use outside the system. Therefore, the accessibility of the original face image is typically the gating factor in preventing use of the face for unforeseen applications in the future.

CONCLUSION

Face recognition is a mature technology capable of searching millions of faces in less than a second on modern computer hardware. The usefulness of the technology has been validated over the last decade by several government customers as a tool for fighting identity fraud, crime, and terrorism. Over the next decade, we expect that face recognition will dramatically improve in matching faces from uncontrolled capture environments and additionally improve match efficiency in the controlled cases, further broadening the scenarios where face recognition can be used effectively.

Thank you for the opportunity to address the Subcommittee on these important issues. I look forward to answering your questions.

In: Facial Recognition Technology ... ISBN: 978-1-62417-572-5
Editor: Meena N. Harnois © 2013 Nova Science Publishers, Inc.

Chapter 7

TESTIMONY OF ROBERT SHERMAN, MANAGER OF PRIVACY AND PUBLIC POLICY, FACEBOOK. HEARING ON "WHAT FACIAL RECOGNITION TECHNOLOGY MEANS FOR PRIVACY AND CIVIL LIBERTIES"*

Chairman Franken, Ranking Member Coburn, and Members of the Committee, my name is Robert Sherman, and I am the Manager of Privacy and Public Policy at Facebook. Thank you for the opportunity to share Facebook's views on what the increasing use of facial recognition technology means for the American people. Facebook is committed to building innovative tools that enhance people's online experiences while giving them control over their personal information. Our integration of facial recognition technology into tag suggestions on Facebook exemplifies this commitment, and I look forward to discussing with you the privacy and security protections that we have built into this feature.

At Facebook, we understand the importance of continuing to build innovative technologies that help people communicate and share in a way that honors and preserves the trust that they have placed in us. Indeed, we believe that our success as an innovator is largely due to the work that we do every

* This is an edited, reformatted and augmented version of testimony given on July 18, 2012 before the Senate Judiciary Committee, Subcommittee on Privacy, Technology and the Law.

day to build and maintain people's confidence that we will be good stewards of their data.

In my testimony today, I will first describe the important user controls that we include in our tag suggestions feature. Second, I will address the steps we take to safeguard the data we maintain in connection with tag suggestions.

I. Facebook's Photo Management Tools, Including Tag Suggestions, Were Designed with Privacy at the Forefront

At Facebook, we implement facial recognition technology in our popular and innovative tag suggestions tool, which helps users manage and share their photos. In the early days of Facebook, we learned how important photo sharing was to our users when we realized that people were frequently changing their profile photos to show friends their most recent snapshots. In response to that feedback, in 2005 we built a feature that allowed people to upload and share photos on Facebook, and we have continued building those features ever since.

One component of our photo management and sharing features is photo tagging, which is the 21st century's version of handwriting captions on the backs of photographs, and it allows users to instantaneously link photos from birthdays, vacations, and other important events with the people who participated. To help our users more efficiently tag their friends in photos, we built tag suggestions, which uses facial recognition technology to suggest people they already know and whom they might want to tag.

In recent years, other companies have begun using facial recognition technology as well, and many photo management services incorporate tools similar to ours. Today, photo sharing is so popular on Facebook that as many as 300 million new photos are uploaded to our service each day. To keep up with that demand, a few months ago we took our tag suggestions feature down to improve its efficiency. We plan to restore the feature to our site soon.

To understand how tag suggestions work, and how we designed the feature to protect our users' privacy, it helps to have a general understanding of the core tools and controls we provide to help people organize and share their photos.

A. Photos and Tagging on Facebook

Facebook's photo management tools are among the most popular features on our service in large part because they build upon things people could always do with their photos — such as placing them in albums and adding captions — by making photos more social. The key to making photos social on Facebook is our "tagging" feature. On Facebook, a "tag" is a special link that can be used to associate a photo with a particular user's page on Facebook, which we call a "timeline," and, depending on the person's privacy settings, share that photo with others. The popularity of tagging is no doubt due to the social interaction it encourages. Tags also help promote transparency and control on Facebook because when a person is tagged, by default Facebook lets that person know. This allows the person included in the photo to interact with the user who uploaded it or to take action if he or she does not like the photo, such as removing the tag or requesting that the photo be removed entirely. And the photo may be shared with others (again, depending on privacy settings), who may "Like" or comment on the photo.

Another reason why photos and tagging are such popular features on Facebook is the fine-grained privacy controls built into those tools. For example:

- Inline privacy controls. When people share photos on Facebook, our inline audience selectors enable them to determine with precision the audience with whom the photos will be shared. These controls are presented at a time and in a way that allows a person to make a meaningful decision about her photos when he or she is posting them.
- Tag review. Tag review allows people to pre-approve stories where they are tagged before they appear on their timeline.
- Tag removal. A person can "untag" a photo in which she appears, thereby unlinking it from her account.
- Blocking. Facebook's privacy settings allow people to "block" others who tag them in photos. Blocking prevents the blocked person from tagging the person again in photos.
- Reporting. Photos on Facebook contain a "Report" link, which enables people to request removal of objectionable photos. Through this link, a person can easily contact the person who posted the photo to make a deletion request or refer the photo to Facebook for professional review.

As with all our products and services, we constantly strive to make our photo management tools more responsive to the needs of people who use Facebook. After we introduced tagging, many people told us that the feature was useful but that manually entering tags for each person in every photo required a great deal of time and effort. We developed tag suggestions in response to this feedback.

B. Tag Suggestions

Tag suggestions automate the process of identifying and, if the user chooses, tagging friends in the photos he or she uploads. Convenience features like this, which are designed to automate tools, are common with online services and often greatly enhance everyone's experience – especially with popular products like photo sharing. Tag suggestions works by determining what several photos in which a person has been tagged have in common and storing a summary of the data derived from this comparison in a file that we call a "template". When a person uploads a new photo, we compare that photo to the summary information in the templates of the people on Facebook with whom the person communicates most frequently. This allows us to make suggestions about whom the user should tag in the photo, which the user can then accept or reject.

Tag suggestions has been enthusiastically embraced by millions of people on our service because it is convenient and the uploader is in control of their photos. We recently acquired Face.com, the company from which we licensed the technology that we use to operate tag suggestions. The engineers who worked at Face.com are working with us to improve the efficiency of our tag suggestion systems. In the interim, we are working to wind down Face.com's operations.

When we first designed tag suggestions, we considered how to implement the technology in a way that would respond to people's demand for better tagging tools without diminishing their ability to control the collection, use, and disclosure of their information. We launched the feature with several important privacy protections.

First, we are transparent about our use of the technology. Our Data Use Policy[1] contains a clear and concise explanation of tag suggestions and links to a series of frequently asked questions in our Help Center,[2] which describes how tag suggestions work and how people can control the feature.

Second, tag suggestions only use data people have voluntarily provided to Facebook — photos and the tags people have applied to them — and derives information from that data to automate the process of future tagging. We do not collect any new information beyond the photos themselves in order for tag suggestions to work. In this regard, tag suggestions are similar to the recommendations offered by popular sites such as Amazon and Netflix. Like those services, tag suggestions simply takes data that people have provided and uses that data to make helpful, time-saving predictions about how people may want to use our site. Unlike implementations of facial recognition that rely on the direct collection of data from people with whom the collector has no preexisting relationship, the people who provide the data that powers tag suggestions have established relationships with Facebook and choose to use our service because they want to share information about themselves in a safe and controlled way.

Third, Facebook's technology does not enable people to identify others with whom they have no relationship. Tag suggestions simply use a list of a person's friends who have been tagged in other photos and suggests which of those friends might be in the photos she uploads. Facebook's technology is not designed to search for random strangers – rather, it is optimized to help you automate the tagging of the people in your photos so you can share those photos with them.

Fourth, and perhaps most importantly, Facebook enables people to prevent the use of their image for facial recognition altogether. Through an easy-to-use privacy setting, people can choose whether we will use our facial recognition technology to suggest that their friends tag them in photos. If a person chooses not to allow this, as we explain on our website: "When you turn off tag suggestions, Facebook won't suggest that friends tag you when photos look like you. The template that we created to enable the tag suggestions feature will also be deleted."[3]

Facebook's tag suggestions feature, unlike other facial recognition products, is based specifically on individual control. It makes suggestions based on the networks that people have formed expressly on Facebook, rather than seeking to identify unknown people. And it further empowers people to control the experience, even within those expressly created networks, by notifying them when they are tagged and giving them a range of choices about how information derived from that tag will be used.

II. FACEBOOK PROVIDES STRONG SAFEGUARDS FOR THE DATA USED FOR TAG SUGGESTIONS

In addition to providing appropriate privacy controls, Facebook has implemented strong technical and procedural safeguards for the personal information we store, including the templates used to power tag suggestions. We encrypt these templates as they are stored on our servers and limit the ability of people within Facebook to access this information.

In addition, the nature of our technology, federal law, and our internal policies restrict the facial recognition data we disclose to third parties, including law enforcement. Two aspects of our unique technology severely limit its usefulness to law enforcement agencies. First, our templates work only with our proprietary software. Alone, the templates are useless bits of data. Second, our software cannot be used to compare a photo of an unknown person against our database of user templates. Our technology is designed to search only a limited group of templates — namely, an individual user's friends — and law enforcement agencies accordingly cannot use our technology to reliably identify an unknown person.

Moreover, as we describe in our Law Enforcement Guidelines[4] — we are one of only a handful of major Internet companies to post them online — Facebook has implemented rigorous procedures meant to ensure that information is shared with law enforcement only in very limited circumstances pursuant to applicable law. As we describe on our website, we disclose account records solely in accordance with our terms of service and applicable law. Indeed, a dedicated team of professionals scrutinizes each request for legal sufficiency and compliance with Facebook's internal requirements, and we regularly oppose requests that we believe violate the law. Our Data Use Policy also provides the ability to share information with law enforcement in emergency situations including risk of death or bodily harm. Such situations are rare, and we take very seriously the commitments we have made to people who use Facebook in our Data Use Policy.

CONCLUSION

I hope that my testimony today has helped the Members of this Committee understand how Facebook uses facial recognition technology and, more importantly, the privacy and security protections that define our implementation of the technology. The tools we create to enable people to share information with friends and family — and the innovative ways in which we let users control their information — will no doubt continue to evolve. But our fundamental commitment to protect the privacy and security of the people who use Facebook will not change. We look forward to continuing our dialogue with you on this important subject.

Thank you for the opportunity to testify today. I look forward to answering any questions you may have.

End Notes

[1] *See* https://www.facebook.com/full_data_use_policy.
[2] *See* https://www.facebook.com/help/tag-suggestions.
[3] https://www.facebook.com/help/?faq=187272841323203#How-can-I-turn-off-tag-suggestions?
[4] https://www.facebook.com/safety/groups/law/guidelines/.

In: Facial Recognition Technology ...
Editor: Meena N. Harnois

ISBN: 978-1-62417-572-5
© 2013 Nova Science Publishers, Inc.

Chapter 8

Testimony of Maneesha Mithal, Associate Director, Division of Privacy and Identity Protection, Federal Trade Commission. Hearing on "What Facial Recognition Technology Means for Privacy and Civil Liberties"*

I. Introduction

Chairman Franken, Ranking Member Coburn, and members of the Subcommittee, I am Maneesha Mithal, Associate Director of the Division of Privacy and Identity Protection at the Federal Trade Commission ("FTC" or "Commission"). I appreciate the opportunity to present the Commission's testimony on facial recognition technologies.[1]

Facial recognition technologies currently operate across a spectrum, ranging from pure facial detection, which simply means detecting a face in an image, to biometric analysis of facial images, in which unique mathematical data are derived from a face in order to match it to another face.[2] In the latter example, if one of the faces is identified – *i.e.* the name of the individual is

* This is an edited, reformatted and augmented version of testimony given on July 18, 2012 before the Senate Judiciary Committee, Subcommittee on Privacy, Technology and the Law.

known – then in addition to being able to demonstrate a match between two faces, the technology can be used to identify previously anonymous faces. In between these two points are a range of possibilities that include determining the demographic characteristics of a face, such as age range and gender, and recognizing emotions from facial expressions.

Having overcome the high costs and poor accuracy that once stunted their growth, facial recognition technologies are quickly moving out of the realm of science fiction and into the commercial marketplace.[3] Today facial recognition technologies can be found in a wide array of contexts, including digital signs, mobile applications, and social networks. While consumers may enjoy the benefits associated with advancements to these technologies – such as easier organization of online photos – there are also concerns that the technologies may increase the risks to consumer privacy. Recognizing that the commercial use of these technologies will likely continue to grow, the FTC has sought to understand how these technologies are being used, how they could be used, and how they will shape consumers' commercial experiences.

To examine these issues, the FTC hosted a workshop in December 2011 – "Face Facts: A Forum on Facial Recognition Technology" ("Face Facts workshop").[4] Researchers, academics, industry representatives, and consumer and privacy professionals all took part in a series of wide-ranging discussions. Major topics included the recent advances, current uses, and possible future uses of facial recognition technologies, as well as the privacy and security concerns those issues raise. Following the workshop, Commission staff requested public comments regarding a number of topics and questions.[5] Commenters were asked to provide input on, among other issues: the privacy and security concerns surrounding the commercial use of these technologies, best practices for providing consumers with notice and choice about the use of these technologies, and best practices for deploying these technologies in a way that protects consumer privacy. The FTC received eighty public comments from private citizens, industry representatives, trade groups, consumer and privacy advocates, think tanks, and members of Congress, reflecting a wide variety of viewpoints on these issues.[6] We are still reviewing these comments, and staff plans to use the information we have learned to date to release a report later this year setting forth recommended best practices for using facial recognition technologies in a manner that respects consumer privacy while still allowing consumers to receive the benefits these technologies may provide, such as convenience and more personalized service. The report would not serve as a template for law enforcement actions or regulations under laws currently enforced by the FTC.

The FTC is also considering how the three core principles articulated in the Commission's March 2012 report on consumer privacy ("Privacy Report") – privacy by design, simplified consumer choice, and transparency – can be applied to the use of facial recognition technologies.[7] These principles call upon companies handling consumer data to implement privacy by design by building in privacy protections at every stage in the development of their products and services, provide consumers with simplified choices about the collection and use of their information, and increase transparency by providing clearer, shorter and more standardized privacy notices.

This testimony addresses solely commercial uses and does not address the use of facial recognition technologies for security purposes or by law enforcement or government actors. It describes the current facial recognition landscape, including: (1) recent advances in facial recognition technologies, (2) current commercial uses of facial recognition technologies, and (3) possible future commercial uses of facial recognition technologies. The testimony concludes by setting forth some privacy considerations the Commission is examining as staff prepares its facial recognition report and weighs next steps in this area.[8]

II. Current Facial Recognition Landscape

A. Recent Advances in Facial Recognition Technologies

Until recently, because of high costs and limited accuracy, facial recognition technologies were not widely used on a commercial basis. However, recent years have brought steady improvements in these technologies. Several developments have contributed to the increased accuracy in facial recognition systems. For example, better quality digital cameras and lenses create higher quality images, from which biometric data can be more easily extracted. In addition, the goal of some facial recognition technologies is to match an image of an unknown face to an identified "reference photo," where the name of the individual is known. Until recently, it was difficult to match two images if the photos were taken from different angles. With current technologies, companies can generate 3D face images to help reconcile pose variations in different images.

These recent technological advances have been accompanied by rapid growth in the availability of identified photos online. Previously, most of the images available online were of celebrities, but today there are many sources of identified images of private citizens online. One explanation for this is the rise in popularity of social networking sites. For example, approximately 2.5 billion photos are uploaded to Facebook each month.[9] This multitude of identified images online can eliminate the need to purchase proprietary sets of identified images, thereby lowering costs and making facial recognition technologies commercially viable for a broader spectrum of commercial entities.[10]

B. Current Commercial Uses of Facial Recognition Technologies

As noted, facial recognition technologies currently operate across a spectrum ranging from the ability to determine that a photo has a face in it ("pure facial detection") to the ability to identify demographic characteristics of a face, to the ability to match different images of the same face and possibly identify an unknown face. In many cases, the privacy risks associated with commercial uses of facial recognition increase along with the sophistication of the technology at use. For example, the privacy risks associated with companies using facial recognition to identify an unknown face are generally much greater than the risks raised by a company using pure facial detection to locate a face in an image.

Current uses of pure facial detection include, among others, refining search engine results to include only those results that contain a face, locating faces in images in order to blur or de-identify them, or ensuring that the frame for a video chat feed actually includes a face. Pure facial detection is also used in virtual eyeglass fitting systems and virtual makeover tools that allow consumers to "try on" a pair of glasses or a new hairstyle online. In these systems, after the consumer has uploaded a photo of herself to the website, that photo is scanned, basic facial features are picked out and – using the detected facial features as reference points – the eyewear or hairstyle is superimposed on the consumer's face.

More sophisticated technologies that do not merely distinguish a face from surrounding objects, but also assess various characteristics of that face, can be used commercially in a variety of ways. For instance, companies can use technologies that identify moods or emotions from facial expressions to determine a player's engagement with a video game or a viewer's excitement during a movie.

Companies are also using technologies that determine demographic characteristics to deliver targeted advertisements in real-time in retail spaces.[11]

These companies place cameras – which assess the age range and gender of the consumer standing in front of the screen – into digital signs or kiosks. They then display an advertisement based on that consumer's assessed demographic characteristics. For example, a 30 year-old male might be shown an advertisement for shaving cream, while a 50 year-old female may be shown an advertisement for perfume. As currently implemented, companies do not appear to be storing images processed by digital signs for future use.

Digital signage is an area where industry trade groups have proactively issued guidance and "best practices" for their members. For example, Point of Purchase Advertising International's Digital Signage Group ("POPAI") has developed a code of conduct containing recommendations for marketers to follow in order to maintain ethical data collection practices in retail settings.[12] Similarly, the Digital Signage Federation worked with the Center for Democracy and Technology to craft a voluntary set of privacy guidelines for their members, which include advertisers and digital sign operators.[13] Both of these self-regulatory codes address the use of facial recognition technologies in digital signs.

One company has leveraged this ability to determine age range and gender in order to obtain aggregated demographic data about the clientele of bars and nightclubs via cameras placed at the entrance to these venues. This company only stores the aggregated demographic data, and not images of the venues' customers. Both the operators of the venue and third parties – such as liquor distributors – can use this data to understand the demographics of a particular venue's customers at certain times, and possibly tailor their specials or promotions accordingly. This company also makes the aggregate information it collects available through a mobile app that consumers can use to make decisions about which venues to patronize.[14]

Facial recognition technologies that are used to actually identify individuals, rather than simply to detect a face or demographic characteristics, work by deriving unique biometric data from facial images. This biometric data is the unique mathematical characteristics that are extracted from the image in order to capture the individual identity (e.g., distance between eyes, ears, size of features, etc.). Those unique mathematical characteristics can then be compared to the characteristics extracted from other facial images to determine if there is a match.[15]

This type of technology has been implemented in a variety of manners. For example, a mobile phone user can authenticate herself by using her face, rather than a password, to unlock her phone. One of the most prevalent current uses of this technology is to enable semi-automated photo tagging or photo organization on social networks and in photo management applications. On social networks these features typically work by scanning new photos a user uploads against existing "tagged" photos. The social network then identifies the user's "friends"[16] in the new photos so the user can tag them. As currently implemented, these features on social networks suggest "tags" only of people that the user already knows, either through a "friend" relationship or other contacts that suggest the two individuals know each other.

C. Possible Future Commercial Uses of Facial Recognition Technologies

Future uses of facial recognition technologies may provide new and exciting products and services that consumers want. They may also provide privacy and security benefits. For example, as noted above facial recognition technology can be used to authenticate users on mobile devices. In the future, we can foresee broader use of these technologies for authentication purposes which can enhance privacy and security for consumers.

At the same time, there may be privacy and security concerns. For example, will it become feasible to use facial recognition to ·identify previously anonymous individuals in public places, such as streets or retail stores, or in previously unidentified photos online? While it does not seem that it is currently possible for commercial entities to accomplish this on a wide scale, recent studies suggest that in the near future, it may be possible. For example in a 2011 study, Carnegie Mellon researchers were able to identify individuals in previously unidentified photos from a dating site, by using facial recognition technology to match them to their Facebook profile photos.[17]

Some have surmised that advances in facial recognition technologies may end the ability of individuals to remain anonymous in public. If these predictions come to fruition, companies could employ facial recognition technologies in a number of ways that raise significant privacy concerns. For example, companies could match images from digital signs with other information to identify customers by name and target highly-personalized ads to them based on past purchases, or other personal information available about them online. Further, a mobile app that could, in real-time, identify previously anonymous individuals on the street or in a bar could cause serious privacy and physical safety concerns, although such an app might have benefits for some consumers.

III. QUESTIONS AND NEXT STEPS

In its March 2012 Privacy Report the Commission articulated three core principles for companies to consider in protecting consumer privacy:

(1) **Privacy by Design**: The Commission called on companies to build in privacy at every stage of product development. Such protections include providing reasonable security for consumer data, collecting only the data that is consistent with the context of a particular transaction or the consumer's relationship with the business, retaining data only as long as necessary to fulfill the purpose for which it was collected, safely disposing of data no longer being used, and implementing reasonable procedures to promote data accuracy. The Commission also called on companies to implement and enforce procedurally sound privacy practices throughout their organizations, including, for instance, assigning personnel to oversee privacy issues, training employees on privacy issues, and conducting privacy reviews when developing new products and services.

(2) **Simplified Consumer Choice**: The Commission noted that, for practices that are not consistent with the context of a transaction or a consumer's relationship with a business, companies should provide consumers with choices at a relevant time and context. In addition, companies should obtain affirmative consent before (1) collecting sensitive data or (2) using consumer data in a materially different manner than claimed when the data was collected.

(3) **Transparency**: The Commission called on companies to increase the transparency of their data practices so that interested parties can compare data practices and choices across companies. The Commission also suggested that companies – particularly those that do not interact with consumers directly, such as data brokers – provide consumers with reasonable access to the data that the companies maintain about them.

The Commission intends to release a report this year laying out recommended best practices for the use of facial recognition technologies that build on comments by workshop panelists, written submissions, and these three core principles. In developing the report, the Commission is considering the following questions.

First, the Commission is considering ways in which companies using facial recognition technologies can implement "privacy by design." For example, how can companies establish and maintain sound retention and disposal practices for the consumer images and biometric data that they collect? For instance, should digital signs using demographic detection ever store consumers' images? Are there certain sensitive areas where companies should not place digital signs? Are there ways that the use of facial recognition technologies may increase consumer information privacy and security?

Second, the Commission is examining how companies that use these technologies can provide consumers with simplified choices about the collection and use of their data. How should companies using facial recognition technology provide choices? Under what circumstances should companies seek consumers' affirmative express consent before engaging in facial recognition?

Finally, the Commission is considering how companies using facial recognition technologies can increase the transparency of their data practices. For example, are consumers aware that digital signs using demographic detection are being used in retail environments? How can they be made aware? Similarly, how many consumers know that social networks have begun implementing facial recognition for photo "tagging"? How and when should these social networks disclose their practices to consumers? The Commission is currently evaluating these and other questions as it develops a final report on the use of facial recognition technologies in commercial environments.

CONCLUSION

Thank you for the opportunity to provide the Commission's views on the topic of facial recognition. We look forward to continuing to work with Congress and this Subcommittee on this important issue.

End Notes

[1] This written statement represents the views of the Federal Trade Commission. My oral presentation and responses to questions are my own and do not necessarily reflect the views of the Commission or of any Commissioner. Commissioner J. Thomas Rosch dissents to certain portions of the testimony. His views are explained in the attached separate statement.

[2] *See* Dr. Joseph J. Atick, International Biometrics & Identification Association, *Face Recognition in the Era of Cloud and Social Media: Is it Time to Hit the Panic Button?* (Dec. 2011), at 2, *available at* http://www.ibia.org/resources/.

[3] Throughout this testimony, the term "facial recognition" is used broadly to refer to technologies that are used to extract data from facial images. *See* Sony, Face Recognition Technology, http://www.sony.net/SonyInfo/technology/technology/theme/sface01.html.

[4] FTC Workshop, *Face Facts: A Forum on Facial Recognition Technology* (Dec. 8, 2011), http://www.ftc.gov/bcp/workshops/facefacts/.

[5] *See* Press Release, FTC, FTC Seeks Public Comments on Facial Recognition Technology (Dec. 23, 2011), *available at* http://www.ftc.gov/opa/2011/12/facefacts.shtm.

[6] *See* FTC, # 402; FTC Seeks Public Comments on Facial Recognition Technology; Project Number P115406, http://www.ftc.gov/os/comments/facialrecognitiontechnology/index.shtm.

[7] FTC, *Protecting Consumer Privacy in an Era of Rapid Change, Recommendations for Businesses and Policymakers,* (Mar. 2012), *available at* http://www.ftc.gov/os/2012/03/1203 26privacyreport.pdf.

[8] This hearing, and therefore this testimony, focuses specifically on facial recognition technology. However, the Commission is aware that there have also been recent advances in other forms of biometric technologies, such as voice recognition, which may raise similar privacy concerns. Accordingly, the Commission is working to better understand the privacy implications of all forms of biometric technology that commercial entities are using.

[9] *See* Chris Putnam, *Faster Simpler Photo Uploads*, THE FACEBOOK BLOG (Feb. 5, 2010), http://blog.facebook.com/blog.php?post=206178097130.

[10] *See* Center for Democracy & Technology, *Seeing is ID'ing: Facial Recognition & Privacy*, Center for Democracy & Technology (Jan. 22, 2012) at 3, *available at* https://www.cdt.org/files/pdfs/Facial Recognition and Privacy-Center for Democracy and _Technology-January_2012.pdf.

[11] Shan Li and David Sarno, *Advertisers start using facial recognition to tailor pitches*, LA TIMES, Aug. 21, 2011, *available at* http://articles.latimes.com/2011/aug/21/business/la-fi-facial-recognition-20110821.

[12] POPAI, Digital Signage Group, *Best Practices: Recommended Code of Conduct for Consumer Tracking Research* (Feb 2010) *available at* http://www.popai.com/docs/DS/2010 dscc.pdf.

[13] *See* Digital Signage Federation, *Digital Signage Privacy Standards* (Feb. 2011) *available at* http://www.digitalsignagefederation.org/Resources/Documents/Articles%20and%20Whitepapers/DSF%20Digital%20Signage%20Privacy%20Standards%2002-2011%20%283 %29.pdf.

[14] *See* SceneTap, http://www.scenetap.com/.

[15] *See* Dr. Joseph J. Atick, International Biometrics & Identification Association, *Face Recognition in the Era of Cloud and Social Media: Is it Time to Hit the Panic Button?* (Dec. 2011), at 2, *available at* http://www.ibia.org/resources/.

[16] We use the term "friend" to refer to an individual user that another user has a mutual connection with on the social network.

[17] *See Face Recognition Study - FAQ*, http://www.heinz.cmu.edu/~acquisti/ face-recognition-study-FAQ/. This study used a limited geographic area, and therefore a limited number of photos and subjects; thus, the results cannot necessarily be duplicated on larger scale. *See Face Facts Workshop, Remarks of Prof. Alessandro Acquisti, Carnegie Mellon University*, at 130-131 and 138-139.

In: Facial Recognition Technology ... ISBN: 978-1-62417-572-5
Editor: Meena N. Harnois © 2013 Nova Science Publishers, Inc.

Chapter 9

STATEMENT OF JEROME M. PENDER, DEPUTY ASSISTANT DIRECTOR, CRIMINAL JUSTICE INFORMATION SERVICES DIVISION, FEDERAL BUREAU OF INVESTIGATION. HEARING ON "WHAT FACIAL RECOGNITION TECHNOLOGY MEANS FOR PRIVACY AND CIVIL LIBERTIES"*

HISTORY OF THE CRIMINAL JUSTICE INFORMATION SERVICES DIVISION

The Criminal Justice Information Services (CJIS) Division, the largest division in the FBI, was established in February 1992 to serve as the focal point and central repository for criminal justice information services. Collection of fingerprints by the FBI had begun 70 years earlier with the creation of the FBI's Identification Division in 1924. Prior to 1924, states maintained individual repositories of fingerprints and shared information at the state level. It was not until 1924, with the creation of the FBI's Identification Division, that fingerprints were shared on a national level.

* This is an edited, reformatted and augmented version of testimony given on July 18, 2012 before the Senate Judiciary Committee, Subcommittee on Privacy, Technology and the Law.

From 1924 until 1999, fingerprints and associated criminal history information, including mug shot photographs, were received in the U.S. mail and processed manually. In 1999, with the launching of the Integrated Automated Fingerprint Identification System (IAFIS), fingerprints and associated criminal history information were searched, processed, and stored.

Next Generation Identification

The FBI initiated the Next Generation Identification (NGI) Program in response to advances in technology, FBI customer requirements, growing demand for IAFIS services, and growing obsolescence of the IAFIS Information Technology infrastructure. The NGI Program, which is on scope, on schedule, on cost, and 60 percent deployed, is enabling the FBI to meet its criminal justice mission and continue to build its reputation as the global leader in biometrics. The NGI Program is dramatically improving the major features of the current IAFIS, including system flexibility, storage capacity, accuracy and timeliness of responses, and interoperability with other systems. The NGI Program is addressing the increase in identification requests, in the form of biometric submissions, and the rapidly expanding database of biometric information and new biometric identifiers concomitant with evolving tribal, local, state, federal, international, and intelligence requirements. NGI improvements and new capabilities are being introduced across a multi-year time frame through a phased, incremental approach, resulting in a flexible framework of core capabilities that will serve as a platform for multimodal functionality.

General Authority for Next Generation Identification Initiatives

28 U.S.C. § 534 authorizes the FBI to acquire, collect, classify, and preserve identification, criminal identification, crime, and other records. 28 U.S.C. § 534 further enables the exchange of the aforementioned records and information with, and for the official use of, authorized officials of Federal, State, local, and tribal criminal and noncriminal justice departments and agencies.

In addition, 42 U.S.C. § 3771 authorizes the Director of the FBI to develop new or improved approaches, techniques, systems, equipment, and devices to improve and strengthen criminal justice.

STATUS OF NEXT GENERATION IDENTIFICATION INCREMENTAL DEPLOYMENT

NGI is being deployed in 7 separate increments to balance operational needs and technical feasibility. Increment 0 (Advanced Technology Workstations (ATWs)) was completed in March 2010. Increment 1 (the Initial Operating Capability) was completed in February 2011. Increment 1 provided more accurate fingerprint searches, increasing the true match rate to 99.6 percent, and improving support for processing flat and less than 10 fingerprints.

Increment 2 (Repository for Individuals of Special Concern (RISC) and Initial Infrastructure) was completed in August 2011. RISC provides mobile fingerprint identification operations on a national level, in time-critical situations, to assist with the identification of: wanted persons; known or appropriately suspected terrorists; sex offenders; and persons of special interest.

NGI currently has 3 Increments in progress (Increments 3, 4, and 5). Increment 3 (Palm Print Searching & Latent Print Searching) is scheduled to deploy in the spring of 2013. Increment 3 will establish the National Palm Print System, provide enhanced latent fingerprint matching, and provide cascaded searches of incoming transactions against unsolved latent and palm prints.

Increment 4 (Rap Back, Facial, and Scars, Marks, and Tattoo (SMT) Search Capabilities and Migration of Remaining IAFIS Functionality to NGI) is targeted to deploy in the Summer of 2014. Increment 4 will provide a National Rap Back Service for notification of the criminal activity of enrolled individuals and access to a national repository for Facial and SMT searches for investigative purposes. Increment 5 (the Iris Pilot) is scheduled for late Summer or Fall of 2013 and will implement a new iris recognition capability. Increment 6 (Technology Refreshment of the NGI system) is slated for 2014.

Next Generation Identification Success

In addition to increased fingerprint accuracy of 99.6 percent, deployment of Increment 1 (AFIT) has allowed operations to reduce the dependency on a supplemental name check, resulting in a 90 percent (weekly) decrease in the number of manual fingerprint reviews required by CJIS Division service providers.

Since deployment of Increment 2 (RISC), 9 states, representing over 500 agencies, have begun participation in the national service; 10 additional states are in the process of implementing RISC. Over 500 transactions are processed daily with a response time of less than 7 seconds and an average weekly hit rate of 6-10 percent.

Facial Recognition

Next Generation Identification Facial Recognition

NGI Increment 4 includes a new facial recognition system. It was deployed as a pilot in February 2012 and is scheduled for full operational capability in the Summer of 2014. The objective of the NGI Facial Recognition Pilot is to conduct image-based facial recognition searches of the FBI's national repository and provide investigative candidate lists to agencies submitting queries.

The goals of the Facial Recognition Pilot are to test the facial recognition processes, resolve policy and processing issues, solidify privacy protection procedures, and address user concerns.

The Facial Recognition Pilot provides a search of the national repository of photos consisting of criminal mug shots, which were taken at the time of a criminal booking. Only criminal mug shot photos are used to populate the national repository. Query photos and photos obtained from social networking sites, surveillance cameras, and similar sources are not used to populate the national repository.

The national repository is updated as transactions, including enrollments and deletions, are submitted by law enforcement users. The national repository contains approximately 12.8 million searchable frontal photos.

The Facial Recognition Pilot permits authorized law enforcement agencies to submit queries for a facial recognition search of the national repository of mug shots. The national repository can be queried by authorized criminal justice agencies for criminal justice purposes.

Access to the national repository is subject to all rules regarding access to FBI CJIS systems information and subject to dissemination rules for authorized criminal justice agencies. Query requests are processed "lights out" (without human intervention), and the results are returned to the submitting agency as an investigative lead in the form of a ranked candidate list.

The investigative response provided to a submitting agency will include the number of candidates requested, in ranked order. The FBI Number/Universal Control Number of each candidate will also be returned, along with a caveat noting that the response should only be used as an investigative lead.

Upon receipt of an investigative response from the FBI, the submitting agency will be responsible for conducting a full investigation of potential matching candidates.

Facial Recognition Privacy Documentation

In accordance with Section 208 of the E-Government Act of 2002, facial recognition was initially addressed by the FBI's June 9, 2008 Interstate Photo System Privacy Impact Assessment (PIA). In coordination with the FBI's Office of the General Counsel, the 2008 Interstate Photo System PIA is currently in the process of being renewed by way of Privacy Threshold Analysis (PTA), with an emphasis on Facial Recognition.

An updated PIA is planned and will address all evolutionary changes since the preparation of the 2008 IPS PIA.

Participating States and Agencies

Initial pilot participants are states or agencies that already have established facial recognition systems. Following completion of the pilot and implementation of the full facial recognition operating capability, additional federal, state, local, and tribal partners and agencies, with or without established facial recognition systems, will be eligible for participation.

Appropriate Use of Next Generation Identification Facial Recognition Technology

Searches of the national repository of mug shots are subject to all rules regarding access to FBI CJIS systems information (28 U.S.C. § 534, the FBI security framework, and the CJIS Security Policy) and subject to dissemination rules for authorized criminal justice agencies. Queries submitted for search against the national repository must be from authorized criminal justice agencies for criminal justice purposes. Each participating pilot state or agency is required to execute a Memorandum of Understanding (MOU) that details the purpose, authority, scope, disclosure and use of information, and the security rules and procedures associated with piloting. Pilot participants are advised that all information is treated as "law enforcement sensitive" and protected from unauthorized disclosure. Pilot participants are informed that Information derived from pilot search requests and resulting responses is to be used only as an investigative lead. Results are not to be considered as positive identifications.

Current Facial Recognition Pilot Participants

In February 2012, the State of Michigan successfully completed an end-to-end Facial Recognition Pilot transaction and is currently submitting facial recognition searches to CJIS. MOUs have also been executed with Hawaii and Maryland; and South Carolina, Ohio, and New Mexico are engaged in the MOU review process for Facial Recognition Pilot participation. Kansas, Arizona, Tennessee, Nebraska, and Missouri are also interested in Facial Recognition Pilot participation.

Conclusion

The FBI's Next Generation Identification Program is on scope, on schedule, on cost, and sixty percent deployed. The Facial Recognition Pilot which began operation in February 2012 searches criminal mug shots and provides investigative leads. The Facial Recognition Pilot is evaluating and solidifying policies, procedures, and privacy protections. Full operational capability for facial recognition is scheduled for the summer of 2014.

In: Facial Recognition Technology ... ISBN: 978-1-62417-572-5
Editor: Meena N. Harnois © 2013 Nova Science Publishers, Inc.

Chapter 10

TESTIMONY OF SHERIFF LARRY AMERSON, PRESIDENT, NATIONAL SHERIFFS' ASSOCIATION. HEARING ON "WHAT FACIAL RECOGNITION TECHNOLOGY MEANS FOR PRIVACY AND CIVIL LIBERTIES"*

Mr. Chairman, Senator Coburn, and Members of the Subcommittee:

Thank you for inviting me to testify today on behalf of the National Sheriff's Association. Chartered in 1940, the National Sheriffs' Association is a professional association dedicated to serving the Office of Sheriff and its affiliates through law enforcement education, training, and information resources. NSA represents thousands of sheriffs, their deputies and other law enforcement professionals, and concerned citizens nationwide.

I applaud the Subcommittee for holding this important hearing on the implications of facial recognition for privacy and civil liberties. These are critical concerns that rightfully need to be debated and the rights of innocent citizens protected from unwarranted interference in their privacy or everyday lives.

On the other hand, advances in technology, and especially facial recognition, which has already been implemented in law enforcement, national defense and the fight against terrorism, are a critical tool in protecting the

---

* This is an edited, reformatted and augmented version of testimony given on July 18, 2012 before the Senate Judiciary Committee, Subcommittee on Privacy, Technology and the Law.

rights of citizens, in ensuring the accurate identification of suspects, prisoners and potential terrorists is almost immediately ascertained, while protecting the safety of our citizens and law enforcement officers.

There is a critical balance between protecting the rights of law abiding citizens and providing law enforcement agencies with the most advanced tools to combat crime, properly identify suspects, catalog those incarcerated in prisons and jails, and in defending America from acts of terrorism.

Most importantly, advances in facial recognition technology over the last 10 years will result in the end of the total reliance on fingerprinting, where it takes hours and days to identify a suspect, fugitive or person being booked into a jail, to the immediate identification of those known to have criminal records, or who are wanted by law enforcement. It will surprise many in the room today to know that there is no national database of those incarcerated in America's jails at any one time. The use of facial recognition to provide instant identification of those incarcerated or under arrest will eliminate many problems while protecting innocent civilians and law enforcement officers.

For instance, utilizing facial recognition in law enforcement would:

- Interconnect law enforcement and Intel organizations to instantly share vital information with accurate identification results.
- Establish a national database of those incarcerated present and past, fugitives, wanted felons, and persons of interest among all law enforcement agencies.
- Allow officers to quickly determine who they are encountering and provide notification if a suspect is wanted or a convicted felon.
- A simple, cost effective, software based solution delivered in Windows based computers with inexpensive non-proprietary off the shelf cameras provide a huge cost savings.
- Demonstrate new capabilities in alias detection, fugitive apprehension, and speed of suspect recognition
- Ensure correct identification prisoners being released and reduce costs associated with conducted administrative procedures.
- Establish a complete national database of incarcerated persons in for the first time in U. S. history no longer could wanted criminals escape detection and arrest due to inefficient processes.

While fingerprints take hours and days for analysis, some advanced facial recognition in use today by U.S. law enforcement, is as accurate as fingerprints but results are obtained in seconds not hours in identifying criminals and perpetrators attempting to use false identities and aliases.

It is also important to point out that facial recognition comes in two forms, 2D and 3D. Only All-aspect 3D Facial systems can protect the privacy of participants who agree to be enrolled, except for in law enforcement or Homeland Security applications. All-aspect 3D cannot search on 2D facial photographs and cannot be invasive of privacy by design. Advanced facial recognition systems remove skin color and facial hair and therefore have no profiling capability.

Currently, the National Sheriffs' Association, Bureau of Prisons and United States Marshalls Service are all in support of utilizing this new three dimensional, holographic imaging technology to eliminate errors in identification; detecting false identities; and immediately identifying dangerous suspects, fugitives or terrorists rather than learning after they are released on traffic offenses or let go without suspicion because immediate identification is not possible.

Accidental releases, sometimes of dangerous felons, would also be eliminated. This technology has been in use for over 8 years in Georgia Detention Facilities with data bases of approximately 5 million inmates without a single erroneous release.

And just last year, a dangerous murderer was released from the District of Columbia jail by switching a wrist band with another inmate. This cannot happen with facial recognition.

In closing, the proper utilization of facial recognition for intelligence or law enforcement uses, can protect civil liberties, save millions of dollars, and instantly identify fugitives, felons, dangerous suspects and save lives.

Thank you Mr. Chairman and I'll be glad to answer any questions you may have.

INDEX

#

21st century, 89, 102

A

abuse, 47, 50, 64
access, 12, 13, 25, 30, 36, 39, 40, 41, 42, 43, 48, 50, 53, 54, 58, 61, 64, 73, 85, 86, 91, 106, 115, 121, 123, 124
accessibility, 100
accountability, 54
advancements, 40, 110
advertisements, viii, 2, 14, 23, 99, 112
advocacy, 14
aerospace, 94
age, viii, 2, 7, 8, 13, 14, 15, 45, 62, 81, 96, 99, 110, 113
agencies, 32, 36, 38, 39, 40, 41, 48, 52, 57, 81, 85, 106, 120, 122, 123, 124, 126
algorithm, 60, 95, 96, 98
anonymous faces, viii, 2, 7, 110
arrest, 60, 126
assault, 75
assessment, 61
assets, 55, 56, 57, 58, 62, 66
asylum, 57
Attorney General, 50
authentication, 8, 9, 114
authorities, 46
authority, 19, 124
automate, 104, 105
awareness, 19, 31, 68

B

background information, 5
ban, 22
bankruptcy, 23
base, 50
behavioral sciences, 89
benchmarking, 97
benefits, 4, 5, 7, 9, 19, 20, 21, 24, 25, 44, 49, 80, 88, 110, 114
blogger, 90
blogs, 55, 57, 60, 76
blood, 69
buffalo, 66
business model, 19
businesses, 16, 20, 25

C

calculus, 73
candidates, 99, 123
case law, 31, 51
case studies, 5, 11

case study, 24
CBP, 38, 55
cell phones, 97
certification, 54
CFR, 57
challenges, 45, 46, 47
Chief Justice, 61
childcare, 38
children, 3, 9, 14, 55, 85
circulation, 61
CIS, 38
citizens, 6, 48, 61, 112, 125, 126
citizenship, 38, 43, 60
civil liberties, 29, 31, 36, 37, 67, 80, 125, 127
classification, 96
clothing, 69
CNN, 64
codes, 11
color, 127
commerce, 64
commercial, vii, 1, 2, 3, 4, 5, 6, 7, 8, 10, 11, 17, 19, 21, 39, 50, 68, 81, 94, 95, 110, 111, 112, 114, 116, 117
Communications Act, 48, 64
communities, 35, 55, 60
compensation, 63
competition, 20, 21, 49, 88, 91
complexity, 96, 97
compliance, 11, 49, 106
computer, 10, 43, 60, 61, 81, 82, 89, 95, 96, 99, 100
computing, 84, 86
conference, 83
confidentiality, 58
Congress, 2, 20, 33, 37, 51, 75, 80, 110, 116
consent, 3, 9, 13, 15, 17, 18, 21, 25, 30, 42, 47, 49, 50, 53, 64, 85, 87, 88, 115, 116
Constitution, 37, 77
consumer choice, 111
consumer education, 10
consumer protection, 14, 19
consumer surplus, 86
consumers, vii, 2, 3, 4, 5, 7, 8, 9, 10, 12, 14, 15, 17, 18, 19, 20, 21, 24, 25, 26, 33, 49, 88, 91, 110, 111, 112, 113, 114, 115, 116
controversial, 39
convergence, 80, 82
conversations, 73, 78
coordination, 123
cosmetic, 91
cost, 20, 31, 87, 120, 124, 126
cost saving, 126
cost-benefit analysis, 20
covering, 78
CPU, 96
crimes, 33, 44, 98
criminal activity, 121
criminal investigations, 95
criminals, 31, 41, 126, 127
crowds, 31
customers, 8, 9, 24, 25, 36, 100, 113, 114
Customs and Border Protection, 38

D

damages, 50
data collection, 11, 36, 43, 44, 45, 48, 54, 113
data mining, 80, 82, 83
data set, 52, 62
database, 9, 15, 16, 25, 26, 30, 31, 32, 36, 37, 38, 39, 40, 42, 44, 45, 47, 51, 54, 55, 56, 57, 58, 61, 68, 69, 71, 75, 83, 84, 87, 98, 100, 106, 120, 126
demographic characteristics, viii, 2, 6, 7, 13, 15, 110, 112, 113
demographic data, 83, 113
Department of Defense, 36, 41
Department of Homeland Security, 36, 57
Department of Justice, 31, 36, 60
depth, 25

detection, vii, 1, 3, 7, 10, 12, 13, 14, 22, 24, 25, 58, 59, 81, 87, 94, 95, 109, 112, 116, 126
detention, 46, 62, 63, 68, 72
DHS, 36, 38, 40, 41, 55, 56, 57, 58, 60, 62, 65
digital cameras, 6, 111
disclosure, 48, 49, 51, 79, 88, 104, 124
discomfort, 85
distortions, 53
District of Columbia, 127
DNA, 38, 55, 63, 66, 69
DOJ, 31, 36, 40, 56, 57, 58, 60, 62
drugs, 76
due process, 63

E

eavesdropping, 78
economic damage, 85
economics, 79
education, 11, 125
E-Government Act, 123
electronic communications, 48
emergency, 106
employees, 38, 94, 115
employment, 44, 60
encryption, 16
end-users, 84, 86, 88, 90
enforcement, 21, 32, 36, 38, 39, 40, 41, 50, 51, 68, 106, 125, 126
environment, 45, 97
environments, 98, 100, 116
equipment, 121
ethnicity, 45
European Parliament, 61
European Union, 21
evidence, 43, 51, 62, 63, 69, 70, 73, 77
evil, 77
evolution, 87
Executive Order, 57
exercise, 25
extraction, 86, 96

F

Facebook, vi, viii, 7, 8, 16, 18, 22, 23, 24, 26, 30, 32, 36, 39, 42, 43, 46, 47, 48, 49, 50, 55, 58, 59, 60, 64, 81, 82, 83, 86, 88, 89, 90, 91, 101, 102, 103, 104, 105, 106, 112, 114
facial expression, vii, 2, 7, 110, 112
facial recognition technologies, vii, 1, 2, 3, 4, 5, 6, 7, 8, 9, 10, 11, 13, 14, 16, 17, 19, 21, 22, 24, 25, 67, 109, 110, 111, 112, 113, 114, 115, 116
false alarms, 62
false positive, 45, 46, 84, 98
family members, 43
FBI, 31, 32, 36, 38, 39, 41, 45, 55, 56, 57, 58, 60, 61, 62, 78, 93, 119, 120, 121, 122, 123, 124
fear, 31
Federal Bureau of Investigation, viii, 36, 62
Federal Bureau of Investigation (FBI), 36
federal government, 41, 50, 51, 65, 94
federal law, 50, 65, 106
Federal Trade Commission Act, 19, 26, 48
felon, 126
fingerprints, 21, 30, 38, 41, 54, 55, 56, 62, 69, 119, 121, 126
First Amendment, 31
flexibility, 120
food, 44
force, 88
Ford, 26, 57
Fourth Amendment, 31, 37, 46, 47, 51, 62, 63, 64, 65, 67, 68, 69, 70, 71, 72, 73, 74, 75, 76, 77, 78
France, 61
fraud, 30, 41, 94, 95, 98, 100
free association, 37
freedom, 86
fusion, 44

G

GAO, 61
garbage, 63, 76
geometry, 6, 49
Georgia, 127
Germany, 89
glasses, vii, 2, 7, 12, 23, 85, 112
google, 58, 64, 90
governments, 38, 41, 75, 77, 85
GPS, 43, 47, 74, 78
greenhouse, 76
grouping, 42
growth, vii, 1, 6, 19, 110, 111
guardian, 90
guidance, 4, 5, 11, 113
guidelines, 11, 107, 113
guilt, 64

H

habituation, 88
hacking, 9, 87
hair, 84, 97, 127
handwriting, 102
harvesting, 26
Hawaii, 31, 56, 124
health, 3, 14
health care, 3, 14
height, 69
history, 45, 58, 87, 88, 120, 126
host, 63
hotel, 85
human, 6, 80, 84, 87, 89, 95, 99, 123
human brain, 95

I

ICE, 41, 89
ID, 31, 58, 117
ideal, 21, 98
identification, 7, 8, 24, 36, 37, 39, 45, 50, 55, 68, 71, 72, 80, 82, 83, 87, 89, 98, 99, 120, 121, 126, 127
identity, 4, 18, 21, 23, 25, 26, 41, 47, 52, 57, 63, 65, 68, 69, 71, 77, 78, 85, 86, 94, 95, 98, 99, 100, 113
illumination, 21
image, viii, 2, 3, 6, 7, 8, 12, 13, 14, 15, 16, 18, 21, 22, 26, 37, 40, 62, 70, 71, 86, 95, 96, 97, 100, 105, 109, 111, 112, 113, 122
imagery, 68, 69
images, vii, 2, 3, 4, 6, 7, 9, 12, 13, 14, 15, 16, 17, 18, 21, 22, 25, 26, 36, 37, 40, 46, 52, 55, 56, 57, 59, 61, 66, 81, 82, 83, 84, 85, 86, 90, 96, 100, 109, 111, 112, 113, 114, 116, 117
immigrants, 58
immigration, 37, 44, 53, 61
Immigration and Customs Enforcement, 38
Immigration and Customs Enforcement (ICE), 38
impact assessment, 54
Impact Assessment, 39, 40, 55, 56, 57, 58, 62, 123
improvements, 6, 111, 120
indexing, 46
India, 94
individuals, viii, 2, 8, 9, 16, 23, 24, 25, 38, 45, 48, 61, 68, 69, 72, 75, 80, 81, 82, 84, 85, 88, 99, 113, 114, 121
industry, 2, 4, 6, 10, 14, 19, 23, 25, 32, 50, 80, 87, 110, 113
industry representatives, 2, 6, 110
inferences, 4, 80, 83
information sharing, 41
information technology, 79, 85
infrastructure, 120
injury, 20, 21, 49
inmates, 127
innovator, 101
INS, 61

integration, 101
intellectual property, 73
intellectual property rights, 73
intelligence, 41, 120, 127
Intelligence Reform and Terrorism Prevention Act, 57
interference, 63, 71, 72, 73, 125
interoperability, 120
intervention, 88, 99, 123
investment, 4
iris, 55, 56, 68, 69, 93, 121
issues, 2, 4, 5, 10, 19, 21, 45, 49, 100, 110, 115, 122

J

judiciary, 23
Judiciary Committee, 29, 35, 67, 79, 93, 101, 109, 119, 125
justification, 46

K

kill, 60

L

landscape, 75, 111
law enforcement, 4, 5, 20, 21, 31, 32, 35, 39, 40, 41, 48, 50, 51, 52, 53, 55, 65, 67, 68, 69, 75, 94, 106, 110, 111, 122, 124, 125, 126, 127
laws, 4, 5, 31, 37, 43, 49, 50, 65, 76, 110
lawyers, 38
lead, 3, 25, 45, 98, 123, 124
learning, 87, 127
legal protection, 37
legislation, 37, 49, 50, 51, 75, 87
light, 46, 47, 84
light conditions, 84
loans, 44
local government, 39
location information, 43
luggage, 85

M

magnitude, 97
majority, 74, 75
man, 60, 95
management, viii, 2, 8, 22, 94, 102, 103, 104, 114
manipulation, 40
manufacturing, 41
marketplace, vii, 1, 110
Maryland, 31, 56, 78, 124
mass, 37, 46, 84
materials, 40, 51
matter, 14, 32, 67, 75, 89
measurements, 3
media, 23, 26, 37, 49, 60, 90
medical, 63
membership, 63
messages, 43
methodology, 82
metropolitan areas, 38
Mexico, 124
Microsoft, 22
military, 38
mission, 120
Missouri, 124
misuse, 37, 45
mobile device, 36, 82, 83, 84, 114
mobile phone, 4, 8, 9, 18, 23, 82, 83, 113
models, 37, 47, 49, 84, 86, 87
Montana, 63
moratorium, 91

N

NCTC, 57
networking, 17
NIJ, 57
North America, 82

NPR, 36, 61
NSA, 125
NSL, 62

O

Obama, 60
OECD, 65, 89, 91
offenders, 98
officials, 120
omission, 49
online advertising, 23
online information, 83
operations, 19, 104, 121, 122
opportunities, 25, 37, 51
opt out, 59
organize, 9, 81, 102
oversight, 44, 54, 64

P

participants, 86, 123, 124, 127
password, 8, 29, 113
pattern recognition, 95
permission, 32
permit, 50, 65
perpetrators, 127
personal identity, 72, 74, 87
PET, 89, 91
photographs, 25, 36, 38, 39, 45, 46, 48, 55, 56, 59, 62, 64, 69, 102, 120, 127
physical characteristics, 68, 72, 77
physical features, 77
platform, 120
police, 33, 40, 55, 57, 60, 68, 69, 70, 71, 72, 73, 74, 76, 77, 81
policy, 3, 13, 17, 25, 26, 58, 86, 87, 88, 90, 107, 122
politics, 62, 90
population, 36, 38, 55, 59, 62, 84
precedents, 33
preparation, 123
President, vi, viii, 125
presumption of innocence, 45
principles, 5, 11, 12, 14, 15, 18, 51, 65, 111, 115
prisoners, 126
prisons, 126
privacy advocates, 2, 110
Privacy Protection Act, 37, 51
private citizens, 2, 6, 31, 110, 112
private information, 29, 43
private sector, 29, 32, 42, 44, 51, 53, 54, 85, 86
probability, 62
probe, 69, 71
professionals, 6, 106, 110, 125
programming, 22
property rights, 70, 74, 77
protection, 16, 20, 32, 46, 62, 63, 69, 71, 72, 77, 79, 122
public safety, 94

Q

query, 40, 71
questioning, 46, 47

R

race, 37, 45
reading, 50
real property, 70, 74, 75
real time, 81, 82, 83, 85
reality, vii, 1, 30, 45, 83, 86
reasoning, 19
recommendations, 11, 12, 13, 14, 19, 20, 24, 105, 113
refugees, 58
regulations, 4, 5, 51, 110
relevance, 11
relief, 48
religion, 46
religious beliefs, 43, 45, 47, 63

rent, 84
reputation, 120
requirements, 4, 5, 51, 53, 61, 87, 98, 106, 120
researchers, 8, 23, 45, 84, 90, 114
resolution, 97
resources, 21, 24, 26, 85, 117, 125
response, 37, 51, 75, 102, 104, 120, 122, 123
response time, 122
retail, viii, 2, 8, 11, 15, 94, 112, 113, 114, 116
retirement, 77
rights, 13, 31, 48, 50, 61, 63, 64, 74, 77, 125, 126
risk, 14, 45, 53, 61, 87, 88, 106
risk assessment, 61
risks, 4, 15, 18, 20, 37, 43, 88, 110, 112
rules, 51, 52, 53, 54, 123, 124

S

safety, 4, 9, 18, 20, 43, 107, 114, 126
science, vii, 1, 22, 30, 36, 110
scope, 36, 39, 52, 84, 120, 124
security, vii, 2, 3, 10, 12, 13, 16, 20, 21, 39, 40, 43, 44, 45, 46, 50, 51, 52, 53, 54, 55, 56, 57, 60, 62, 64, 70, 71, 72, 73, 75, 77, 81, 83, 86, 91, 94, 99, 101, 106, 110, 111, 114, 115, 116, 124
security concerns, vii, 2, 10, 110, 114
security guard, 44
security practices, 16
seizure, 70, 71, 72, 73, 74, 76
self-regulation, 11, 43, 87
Senate, 26, 29, 35, 50, 65, 67, 79, 93, 101, 109, 119, 125
sensitivity, 3
September 11, 41, 45
servers, 26, 106
service provider, 48, 51, 122
services, 3, 4, 5, 9, 19, 81, 82, 85, 86, 88, 102, 104, 105, 111, 114, 115, 119, 120
sex, 121
sex offenders, 121
sexual orientation, 83, 90
shape, 72, 97, 110
showing, 47, 89
signs, viii, 2, 3, 7, 9, 11, 13, 14, 15, 24, 25, 110, 113, 114, 116
silver, 87
skin, 68, 97, 127
social context, 18
social network, viii, 2, 3, 7, 8, 16, 17, 18, 21, 22, 25, 26, 30, 32, 60, 75, 80, 82, 83, 84, 85, 86, 87, 88, 110, 112, 114, 116, 117, 122
social security, 23, 24, 60, 69
Social Security, 44, 61, 80, 82, 83, 90
society, 31, 37, 44, 45, 51, 54, 63, 74, 76, 80, 87
software, 13, 22, 31, 40, 42, 60, 68, 73, 82, 93, 99, 106, 126
solution, 51, 65, 89, 126
speech, 32, 46, 86
Spring, 58
stakeholders, 10
standardization, 45
state, 12, 14, 15, 22, 35, 38, 39, 41, 49, 50, 51, 56, 63, 65, 72, 83, 86, 94, 97, 99, 119, 120, 123, 124
state laws, 65
states, 20, 31, 36, 38, 39, 41, 49, 56, 58, 72, 94, 119, 122, 123
statistics, 14, 24, 90
statute of limitations, 62
statutes, 37
stigma, 44, 60
storage, 12, 48, 50, 120
Sun, 58
Super Bowl, 90
Supreme Court, 33, 46, 62, 63, 64, 68, 73
surveillance, 31, 40, 44, 53, 64, 76, 85, 86, 97, 122
Switzerland, 89

T

tanks, 2, 110
target, 1, 9, 14, 22, 25, 31, 32, 64, 81, 99, 114
taxes, 33
teams, 93
techniques, 68, 72, 80, 83, 95, 121
technological advances, 6, 19, 75, 111
technological change, 20, 51, 65
technologies, vii, 1, 3, 4, 5, 6, 7, 8, 9, 10, 13, 14, 15, 16, 19, 21, 22, 24, 35, 37, 62, 67, 76, 79, 80, 83, 86, 87, 88, 93, 101, 109, 110, 111, 112, 113, 114, 115, 116, 117
teens, 11
telephone, 78
terrorism, 94, 95, 100, 125, 126
terrorist watchlist, 61
terrorists, 62, 121, 126, 127
testing, 40
texture, 68, 97
theft, 25, 41, 85
think tanks, 2, 110
thoughts, 63, 77
threats, 36, 37, 43
time frame, 120
trade, 2, 10, 14, 49, 50, 79, 85, 87, 110, 113
trade groups, 2, 11, 110, 113
trade-off, 79, 87
training, 79, 115, 125
traits, 81
transactions, 49, 54, 121, 122
transmission, 54
transparency, 3, 4, 10, 12, 14, 15, 25, 44, 103, 111, 115, 116
Transportation Security Administration, 38

U

United, 21, 24, 25, 26, 36, 37, 38, 46, 47, 49, 51, 56, 61, 63, 64, 66, 73, 74, 76, 77, 78, 127
United Kingdom, 21, 24, 66
United States, 25, 26, 36, 37, 38, 46, 47, 49, 51, 56, 61, 63, 64, 73, 74, 76, 77, 78, 127
unreasonable searches, 46, 68, 69, 78
USA, vi, viii, 57, 58, 89, 90, 93, 94, 99
USA PATRIOT Act, 57
user data, 64, 91

V

variations, 6, 96, 111
venue, 8, 14, 15, 25, 81, 113
violence, 71
vision, vii, 1, 95
vulnerability, 91

W

wages, 44
walking, 4, 18, 23, 31, 43, 88
Washington, 49, 50, 56
wealth, 89
web, 26, 49
welfare, 85
wiretaps, 31, 33
witnesses, 33
workers, 38, 63
worldwide, 86, 97